PRAISE ~~~
The Scent of Desire

"Fascinating. . . . A serious book, with many whiffs of delight."
—*Washington Post Book World*

"You'll never take your nose for granted again once you've read *The Scent of Desire*." —*USA Today*

"Intriguing. . . . This illuminating book argues convincingly that the sense of smell should never be taken for granted."
—*Publishers Weekly*

"A lively, seductive exploration of what the nose knows. A delightfully unexpected blend of personal anecdotes, pop-cultural erudition, and scientific understanding." —*Kirkus Reviews*

"Filled with intriguing bits of information." —*Weekly Standard*

"Until Rachel Herz told us so, who would have imagined that the aesthetics and physiology of the nose would both be as rich and complicated as those of the eye and mouth and heart? A charming and informative treatise on a vital and overlooked subject." —Adam Gopnik

"This is a spicy perfume of a book, redolent with fascinating facts and provocative hypotheses."
—Steven Pinker, Johnstone Family Professor of Psychology, Harvard University, author of *The Language Instinct, How the Mind Works,* and *The Stuff of Thought*

"In this tantalizing and evocative book, Herz shows how the sense of smell works in every aspect of our lives, and she awakens our own senses as we read. She makes a complex science accessible, charming, and accurate, and the book is fascinating, witty, and wise."

—Linda Bartoshuk, Center for Smell and Taste,
University of Florida

"Fascinating. The information within the pages of *The Scent of Desire* is astounding and sometimes mind-boggling, but Herz conveys this information well with her wonderfully charming voice and writing style." —Blogcritics.org

"Charming. A reminder that life without aromas . . . would be sad indeed." —Curledup.com

Christine Acebo

About the Author

RACHEL HERZ, PH.D., has been recognized as the world's leading expert on the psychology of smell. Since 2000, she has been on the faculty of Brown University. Her prominence as one of the foremost experts on the sense of smell has led to numerous radio and television interviews, including appearances on the Discovery Channel, ABC News, the BBC, National Public Radio, and the Learning Channel. Examples of her work have appeared in science museums around the country, and she was the subject of a *Scientific American* profile piece. Herz is regularly interviewed in a wide array of print media, ranging from the *New Yorker* and the *New York Times* to *Time* magazine and *O, the Oprah Magazine*.

The
SCENT
of
DESIRE

DISCOVERING OUR

ENIGMATIC SENSE

OF SMELL

Rachel Herz

HARPER ● PERENNIAL

NEW YORK ● LONDON ● TORONTO ● SYDNEY ● NEW DELHI ● AUCKLAND

This book is dedicated to my mother,

Judith Scherer Herz

HARPER ● PERENNIAL

A hardcover edition of this book was published in 2007 by William Morrow, an imprint of HarperCollins Publishers.

HarperCollins books may be purchased for educational, business, or sales promotional use. For information please write: Special Markets Department, HarperCollins Publishers, 10 East 53rd Street, New York, NY 10022.

FIRST HARPER PERENNIAL EDITION PUBLISHED 2008.

Library of Congress Cataloging-in-Publication Data is available upon request.

ISBN 978-0-06-082538-6 (pbk.)

08 09 10 11 12 WBC/RRD 10 9 8 7 6 5 4 3 2 1

CONTENTS

PREFACE

I would like to tell you that my fascination with the sense of smell began at the age of five when I was pulled out of bed one Sunday morning by the aroma of French toast and maple syrup. The facts, however, are somewhat different.

When I was five years old we went for a drive one summer afternoon. The sun was shining, the cicadas were singing, everyone was in a good mood, and the wind blew warm against my face. I remember staring out the car window into blurs of green and yellow grassy fields with the outlines of tall trees pressed against a big blue sky, when suddenly I smelled something strange and pungent. Almost at that exact moment I heard my mother exclaim, "I love that smell!" "What is it?" I asked her. "Skunk," she said with a smile. And from that day on I loved the smell of skunk. I did not realize this was abnormal until a year or so later, when, playing in the schoolyard, I told some friends of my predilection for this scent. I expected nods and maybe some stories about bucolic adventures, but instead my innocent comment was met with derision. "You're weird!" "That's gross!" they chorused, as they ran away.

Why did my schoolmates think this scent was awful and I didn't? I quickly learned that most people shared my schoolmates' reaction and that it was best not to admit this particular scent preference. But to my relief, when I got older I discovered a number of other self-confessed skunk scent lovers like myself. Why do I and a bashful minority like a smell that most people find foul? What makes me different from you this way? Or maybe you are also like me. Why do of us like a particular scent and not another? One of the many things you will discover in this book is why you do or don't like the various scents in your world.

Even before my childhood car ride, where I discovered the joy of skunk aroma, I was fascinated by smell and indeed all things "sensual." I have never been satisfied with merely looking at things but am always impelled to touch and smell objects, at least when it is socially acceptable. I have also always been very sensitive to sounds, many of which I adore, but loud noises are often too much for me, and I only like fireworks displays from a distance. Indeed, watching dancing, exploding lights; the stars; mountains; and the green-blue sea—looking at nearly everything—is among my sensual passions. And of course I am ravenous for the pleasures of food. A few epicurean friends and I have formed and dubbed ourselves the "savor the flavor" club. But, above all of the other senses, smell has always held a special allure and been a mystery to me. Not only is scent beguiling to me and probably to you, too, it is also one of the few remaining scientific enigmas. Many aspects of our sense of smell are still inscrutable to science, and in my professional career exploring its psychological riddles has been my main objective.

My goal to understand the puzzles of smell had been gathering energy ever since I pondered why my scent preferences were not exactly like those of others, but I didn't realize it until many years later. One of the pivotal kernels in laying the foundation for my future endeavors occurred when I was an undergraduate student. It was my last year of college and I was studying for the Graduate Record Exam (GRE). I came across a statement in the general psychology section of the study guide that indicated that the answer to a particular question was that *smells were the best cues to memory.* "Why?" I wondered. "Is this really true?" I asked my honors adviser what he knew about this. He confessed to not knowing why or if this was exactly true, but he had accepted the received wisdom and believed it to be based on brain interconnections. I tucked this curiosity into the pseudo-conscious corner of my brain, where it waited to be awakened several years later.

That moment of awakening came during a course I took in graduate school. The course was a social psychology seminar, in which I came across an article in which the researchers had chosen a surprisingly innovative way to manipulate mood. Rather than the standard verbal or imagery techniques, they had chosen odors: a pleasant odor, almond, for good mood, and the distinctively sour odor, pyridine, for bad mood.[1] The researchers argued that smell was fundamentally connected to emotion and that there was biological and evolutionary evidence underlying this connection. Their evolutionary argument was particularly compelling to me, because before ever delving into scent I was lucky to have made the mistake of studying animal behavior. A mistake for me, that is, because my heart wasn't in it. But what I did learn during this period, and what I then became a devotee of, was evolutionary theory. Evolutionary theory has been a guiding light in my search for meaning and direction regarding the mysteries of smell. You will find evolutionary theory and biological approaches mentioned throughout the pages of this book.

My discovery of this article with its evolutionary arguments and the intriguing use of scent to elicit emotion was a stroke of good fortune, and it struck a deep and resonating chord. I suddenly saw a light at the end of the tunnel, and the days of searching for a dissertation topic to be truly inspired by were resolved in that moment. Thus began my first olfactory mission—to find out whether that GRE statement was really true. *Are odors the best triggers for memories? And what does that mean anyway?* It was during this time that I began to unravel the connections between scent and emotion and memory.

THE DECISION TO STUDY SCENT was not an easy academic path to pursue, because academia, embedded in culture as it is, until very recently did not deem smell to be an important topic to study, just as most of us ignore and take for granted this fifth sense. Fortunately, during my graduate student days, I had an open-minded dissertation adviser, and one of the most important things he taught me was that no one person or theory has the complete answer to any question. This has been an essential position to take for studying the sense of smell. Unlike the other senses, whose basic mechanisms have already been exposed, the Rosetta stone of olfaction has yet to be revealed, and will only be fully realized with a convergence of multidisciplinary methods and theoretical approaches.

It was also during my graduate student days at the University of Toronto that I read Diane Ackerman's gorgeously written *Natural History of the Senses*. I felt vindicated that, while the ivory tower didn't give much credence to the intrigues of olfaction, Ackerman devoted a fifth of her book to this marvelous sensory system; her passionate and beautiful prose planted the seed in my mind that one day I would like to share my journey with the sense of smell with others.

Without too much fumbling in the dark, my dissertation quest into the singularities of scent-evoked memories was a success, and you will learn of its results in the pages to come. I then traveled westward to a postdoctoral position at the University of British Columbia where, among other things, I discovered how odors can be used to help us remember all kinds of information. After my postdoctoral fellowship, I began my faculty career at the Monell Chemical Senses Center in Philadelphia, where I spent six years. I then moved to Brown University in Rhode Island in 2000, where I have been academically housed ever since.

During the nearly two decades of my work in olfaction, I have uncovered how and why we like or dislike scents and what is special and different about memories evoked by fragrances. I have uncovered how emotion is a central and fundamental feature of odor perception, odor learning, and odor memory. I have also looked at how the intimate connection between scent and emotion can influence our behaviors. Other avenues I have investigated include how words can be used to create olfactory illusions, and how scents can ignite—or extinguish—sexual attraction. I will reveal these secrets of olfaction to you in this book.

I have published over fifty original articles in scientific journals and written a number of chapters for college textbooks and academic anthologies. I am also involved in the flavor and food community as a distinguished lecturer for the Institute of Food Technologists, and I have been consulting for the major players in the fragrance and flavor industry since the mid-1990s.

I have been lucky enough to have had the opportunity to become a big fish in a small pond. While the molecular and biochemical study of olfaction has mushroomed over the past decade, with research including the groundbreaking discovery of the gene sequence for olfactory receptors by Linda Buck and Richard Axel in 1991—a discovery that was rewarded by the Nobel Prize in 2004—the psychological study of olfaction, by contrast, has remained a very small but fertile patch of ground, and I have been actively digging into it.

Most of all, I have been lucky enough to discover what a marvel our sense of smell is without having to come to this knowledge through tragedy. Several years ago I met a woman who had recently lost her sense of smell in a brutal car accident. You will meet her in the following chapters as Jessica Ross, and her story is part of the inspiration for writing this book. She, like most people, had taken her sense of smell for granted. After losing it, she discovered how unbelievably connected smell had been to everything in her life, and how her life was irrevocably altered and diminished without it.

If you are like most people, you probably haven't given your sense of smell much thought. In a questionnaire study designed by Paul Rozin at the University of Pennsylvania, it was found that when asked to rank the value of losing various physical attributes, the sense of smell was at the bottom. People ranked the loss of their big toe as equivalent to losing their sense of smell. This is a big mistake. The sense of smell, as I'll show, is pivotal to the most important dimensions of our lives.

To know how vital the sense of smell is, we must consider what happens to our lives without it. For those with this devastating condition, called anosmia, *everything* changes. Our sense of smell is essential to our humanity, emotionally, physically, sexually, and socially. Without a sense of smell, our ability to know ourselves and others is obscured; our emotional world becomes deadened or disturbed; our ability to enjoy food is lost; our health may decline; and our sexual desire, and indeed our capability to identify with whom it would be biologically best to conceive a child, is severely weakened. And smell loss is not so rare; it only seems so because it is underreported and the medical response is typically indifferent. From patient data it is conservatively estimated that one in twenty Americans suffers from smell loss, but many people with smell disorders do not see their doctors. In reality the incidence of olfactory dysfunction is shockingly high, and it is one of the most common ravages of aging. Whether it occurs in an instant or gradually over many years, its absence affects those afflicted in both mundane and dramatic ways for the rest of their lives.

Olfaction is the orphan cousin of the senses. Hundreds of books have been written about vision. Library shelves are also full of tomes on the senses of hearing and touch, and books on taste are well stocked because of taste's connection to food. But until recently there have been almost no books about the sense of smell, especially those aimed at the reader who comes to the page with no specialty other than curiosity. My wish is that everyone who reads this book will cease to take their sense of smell for granted. My wish is that from reading this book you will forever after truly make time to stop to smell the roses, or the bacon, and relish all the wonders and gifts that the sense of smell has to offer.

In this book you will discover how the sense of smell has profound influence over how we act and react, taste and think, and behave and love. You will be surprised by how it influences our mental and physical health, what we see in the cinema, and what we buy at the store. You will discover the truths and myths behind aromatherapy and pheromones. You will learn how dogs are saving our lives with their noses and what futuristic innovations, such as "Smell-O-Vision" and "wasp hounds," are waiting in the wings. You will also be entertained and informed by many amazing and useful facts—we can't smell while we are asleep, and without the sense of smell you can't taste the difference between an apple and potato. This book is sprinkled with anecdotes that illustrate breakthrough discoveries in olfaction. For example, in an experiment conducted by the psychologist Michael O'Mahoney, radio listeners were told that a certain sound frequency they could not hear would nevertheless stimulate their sense of smell. After the broadcast, many listeners called the station and complained of strange scents and even allergies, thus illustrating how easily our sense of smell can be tricked and duped by suggestion, and showing one of the ways in which odor illusions can be created.

Our sense of smell is integrally tied to our emotions, our memories, our behaviors, and our health. Scents influence our social relationships and family ties, and they fuel our passions for people and food. From reading this book you will not only have a much greater appreciation and understanding of your astonishing sense of smell, but will also know how to use it to improve and enrich your life in diverse ways. Recognizing how remarkable and valuable your sense of smell is will give you a profoundly sensual and deeper experience of your own existence. It is my privilege to share with you what I have learned about our most enigmatic and emotionally connected sense, our sense of smell—the amazing *sense of desire*.

ACKNOWLEDGMENTS

The origins of this book are the origins of my entrance into the world of smell. My first order of thanks goes to my Ph.D. supervisor, Gerald Chupchick, who enabled me to pursue this unusual avenue of research, and to the mentors I had during my early scientific training—Fergus Craik, Allison Fleming, and Eric Eich. Since that time there have been many scientists, colleagues, friends, and family whose influence and work are referred to throughout this book and without whose input this book would not exist. Most especially, I would like to honor the brilliance and originality of Trygg Engen, my olfactory mentor and the father of the psychological study of olfaction.

My appreciation and dependence upon others for the actual accomplishment of this book extends to many. My former Monell Chemical Senses Center colleagues whom I called upon with various questions, queries, and requests for data and who generously offered their help are Alexander Bachmanov, Beverly Cowart, Julie Mennella, Marcia Pelchat, George Preti, Nancy Rawson, Leslie Stein, Chuck Wysocki, Kunio Yamazaki, and above all Gary Beauchamp.

Other colleagues from both academia and industry whose insights I sought include: John Kauer, Richard Doty, Estelle Campenni, Valerie Duffy, John Hayes, Dana Small, Paul Rozin, Susan Knasko, Linda Bartoshuk, Craig Warren, Stephen Warrenburg, Mark Peltier, and Eileen Kenney. Special appreciation goes to Terry Molnar of the Sense of Smell Institute and The Fragrance Foundation for all of her help with "fine fragrance" related facts, and the Foundation's interest in and support of my work over many years.

I am indebted to various journalists and writers who directly or indirectly influenced my thinking and the development of this book; they are Anne Mullins, Tina Headley, Gabrielle Glaser, Richard Dawkins, Steven Pinker, and Diane Ackerman. I would also like to thank Graig Donini and Sinauer Associates Publishers for permission to use the figure shown in Chapter 1.

Many people facilitated in numerous invaluable ways in this book's creation with their professional assistance and innate talents. For general scholarly and writing advice, particular credit goes to Harriet Bell, John McCann, Kathleen McCann, Barbara Hano, and Nathaniel Herz. I owe my heartfelt gratitude to all the people who acted as silent research assistants, alerting me to the various latest and greatest finds in olfaction: Judith Herz, Jeremy Wolfe, Rachel Tyndale, Mary Carskadon, Eryl Sharp, and Ed Brown. Special appreciation goes to my quick and clever editor, Marjorie Braman, whom I have been lucky enough to work with, and my superb and industrious agent, Wendy Strothman, without whom this book would not have happened, as well as the rest of the staff at the Strothman Agency.

I am extremely grateful to my dearest friends and relations; you were my muses and you buoyed me through the evolution and achievement of this book. A very special thank-you goes to Eryl Sharp, for unflagging encouragement and support, and to Molly, my springer spaniel, whose sense of smell and obsession with scent are a constant source of awe and inspiration.

CHAPTER 1

THE SENSE OF DESIRE

Smells are surer than sights or sounds to make your heartstrings crack.

—NABOKOV

On November 22, 1997, Michael Hutchence, lead singer for the internationally renowned Australian band INXS, was found hanging naked in his hotel bedroom, the noose tied with his own leather belt. The last person to see him alive, onetime fling and longtime friend Kym Wilson, was at first suspected to be complicit. She insisted that she and her boyfriend had left Michael fully clothed the night before after a visit in his hotel room. Kym was soon relieved of suspicion and suicide confirmed as the mode of death. But why would the excessively successful and apparently want-free Michael Hutchence take his own life? What could have drawn him into such a deep depression that he would ultimately kill himself? Various accounts by friends and associates, including interviews with Michael Hutchence himself, point to a pivotal and life-altering event that could very well be the precipitating link to his suicide.

In September 1992, Michael Hutchence was in a freak traffic accident. Riding his bicycle home from a nightclub in Copenhagen, he was struck by a car and suffered a fractured skull. In an interview a few months after Michael's death, the journalist Robert Milliken reported in a feature in *The Independent* (March 1998), that: "His friends are convinced that the accident was a turning point that led to increasing bouts of depression and reliance on Prozac." Richard Lowenstein, the avant-garde Australian filmmaker, told Milliken that ever since the accident Michael was on a slow decline. He had never seen any evidence of depression, erratic behavior, or violent temper before, but saw all those things afterward, and he confessed that one night in Melbourne, Michael had broken down in his arms and sobbed: "I can't even taste my girlfriend anymore."

What was it about this accident that so scarred Michael Hutchence? Did he suffer undiscovered brain damage with pathological effects, or was it something more basic and obvious? Michael was a devout hedonist and completely sensual being.[1] A self-confessed decadent, Michael's gourmand tastes and lust for life were centered around consumption, and now these lascivious pleasures were irrevocably altered, because the accident had stolen his sense of smell.* Without the sense of smell, the temptations of food, the sweaty funk of sex, the

---

*Although his disability was described as losing "smell and taste," people frequently believe that they have lost their sense of taste when they have only lost their sense of smell. This is because flavor is predominantly produced by smell not taste per se (see Chapter 7). It is most likely the case that Michael Hutchence only lost his sense of smell.

essence of a walk on the beach, the feeling of nostalgia—the texture of life itself—were robbed from him. From all accounts, after this accident Michael fell into an increasingly debilitating depression from which he never emerged. As his melancholy progressed, he resorted more and more to both prescription and illicit drugs and alcohol, but these mind-numbing devices were in vain. Could it be that losing his sense of smell, which killed his most basic life pleasures, had such a cataclysmic effect on his well-being that he felt life no longer worth living? From all I know about the sense of smell and the consequences of its loss, this could very well be so.

My suspicion that loss of smell, medically referred to as *anosmia* (smell blindness), was a crucial factor in the suicide of Michael Hutchence is based on my insights into neurological, psychological, and clinical evidence. First, the neurological interconnection between the sense of smell (olfaction) and emotion is uniquely intimate. The areas of the brain that process smell and emotion are as intertwined and codependent as any two regions in the brain could possibly be. Smell and emotion are located in the same network of neural structures, called the *limbic system*. The limbic system is the ancient core of the brain, sometimes called the *reptilian brain* because we share it with reptiles, and sometimes called the *rhinencephalon*—literally, the "nose-brain." The key limbic structure to interact with our olfactory center is the *amygdala*. The amygdala is the brain's locus of emotion. Without an amygdala we cannot experience or process emotional experiences, we cannot express our own emotions, and we cannot learn and remember emotional events. Brain imaging studies have shown that when we perceive a scent the amygdala becomes activated, and the more emotional our reaction to the scent, the more intense the activation is. No other sensory system has this kind of privileged and direct access to the part of brain that controls our emotions.

Clinical research on patients who have lost their sense of smell also suggests that Michael Hutchence's anosmia could have led him to suicide. After an acute trauma such as a head injury, which causes anosmia, patients often report a loss of interest in normally pleasurable pursuits, feelings of sadness, loss of appetite, difficulty sleeping, loss of motivation, inability to concentrate, and thoughts of suicide that can turn into action if not treated.[2] These symptoms are all key diagnostics for major depression as described by the *DSM-IV*,[3] the clinician's bible for classifying psychological disorders. The link between smell loss and depressive symptoms is correlational in humans, but cause and effect has been experimentally verified in laboratory animals. Rats who have had their olfactory bulbs surgically removed, and thereby can no longer perceive smells, display physiological and behavioral changes that are strikingly similar to those that occur in depressed people. They stop eating, lie around their cages, and are oblivious to the toys and activities that they normally vigorously enjoy.

Studies of people afflicted with anosmia also indicate that the development of depression is progressive. In one study that contrasted the trauma of being blinded or becoming anosmic after an accident, it was found that those who were blinded initially felt much more traumatized by their loss than those who had lost their sense of smell. But follow-up analyses on the emotional health of these patients one year later showed that the anosmics were faring much more poorly than the blind. The emotional health of anosmic patients typically continues to deteriorate with passing time, in some cases requiring hospitalization and in more tragic cases, such as Michael Hutchence's, ending in suicide.

I have been contacted by people worrying that their anosmia will *cause* them to commit suicide. They have read or heard a story like Michael Hutchence's and they are experiencing symptoms of depression themselves. The first person who asked me about this was Jessica Ross,* a woman I got to know well when I became the expert witness in her claim against an insurance company.

ONE DAY ABOUT TWO YEARS AGO I was working in my office when the telephone rang. The caller introduced himself as Bill Adams, a partner in a well-established law firm in Florida. He explained that they had a case where a woman had lost her sense of smell in a car accident, and they needed a scent expert to determine the validity and extent of her claims. The purpose of his call was to find out whether I would be willing to be an expert witness in this case.

Jessica Ross, twenty-eight years old, recently married and working at an accounting firm, was coming home from a party seated in the backseat of a car, when it collided with a truck on the highway. She was thrown forward and smashed against the

---

*The names and identifying features of individuals in this case have been changed.

front seats and windshield. Her face took the brunt of the impact, particularly the area around her eyes and forehead, and in less than a year she had already undergone three operations to correct the damage. Jessica's cranial fractures were at the level of the eyebrow, which would have severely affected the brain area behind it—the olfactory bulbs and the olfactory cortex, where our sense of smell is processed. The *olfactory bulbs* are two blueberry-shaped and -sized extensions of the brain, one for each nostril. Separating the neurons in our nose from the olfactory bulbs in our brain is a very thin and fragile bone called the *cribriform plate*. The cribriform plate is riddled with thousands of tiny holes through which the ends of the olfactory receptor neurons (called axons) pass through to get into the brain. When there is a violent blow to the front of the head, it knocks the cribriform plate out of alignment, causing the delicate olfactory axons that run through it to be sheared off. Imagine slicing a cobweb with a knife. With the axons cut off, the olfactory nerve is dead and the sense of smell is destroyed. The axons can never regenerate, and smell loss is permanent. Many types of frontal head injuries can easily lead to losing the sense of smell; for example, football players often suffer the same fate after a hard frontal tackle to the face. This is how Michael Hutchence's fractured skull made him anosmic.

After giving me this report, Adams asked if I thought it was possible that the car crash caused Jessica to lose her sense of smell. Indeed I did. He also explained that although her smell loss was apparent to her many months earlier, she was just now discovering how devastating this disability was. Jessica was now seeking to sue the responsible parties for her anosmia and her loss of quality of life. Adams told me that prior to contacting me he had searched through available documentation but had been unable to find any legal precedent or medical validation for the severity of anosmia. The American Medical Association *Guides to the Evaluation of Permanent Impairment* currently gives the loss of smell and taste a value of only 1–5 percent of the total value of a person's life's worth, while loss of vision is given a value of 85 percent. In spite of this disregard for the importance of the sense of smell, could I help to justify how significant Jessica's disability was? Could it actually be comparable to losing sight?

When her lawyer called me, Jessica was cosmetically put back together. Her doctors had reported no central brain damage, and she was declared of sound mind and body—except for her sense of smell. To confirm her complaints, Jessica had been given a smell identification test* and received a score indicating pure anosmia. There was no doubt in my mind that her smell loss was total and permanent. I asked Adams to arrange a telephone interview so that I could find out how Jessica was coping.

For several hours I bombarded Jessica with a stream of questions concerning all aspects of her life. Jessica told me she felt disconnected from other people and, worse yet, disconnected from her *self*. She believed that she was now incompetent as a homemaker and caretaker, she had lost interest in sexual intimacy, and formerly gregarious, she now avoided

---

*The "UPSIT" (The University of Pennsylvania Smell Identification Test), a scratch-and-sniff odor identification test that is widely used in medical diagnoses of anosmia.

social contact as much as possible. More than anything else, Jessica complained that her emotional life had taken a markedly negative turn since the accident and that she felt generally depressed. "My husband says my personality has changed," she confided. When I asked her to explain these changes, she said that since the accident she was more irritable, was more ambivalent about other people and cared less what others thought of her, and that she felt sad a lot of the time and cried frequently. And haunting her through all of this was a growing worry that at any moment there might be something terribly wrong and that she wouldn't know it because she couldn't detect it as she might have done before with her nose: fire, spoiled food, and even her own body odor.

Throughout our conversation Jessica's voice was flat and dull. Her symptoms and her voice displayed all the classic characteristics of depression. Jessica was also distressed because she felt that her emotional state was getting worse with time. Unfortunately, Jessica was right. The derailing of our olfactory system caused by anosmia can have a progressively negative downstream effect on the healthy functioning of our emotional system.

THE DEPRESSION-OLFACTION LOOP

Jessica Ross, Michael Hutchence, and clinical studies have shown us that anosmia can lead to depression. Conversely, it also turns out, depression can lead to loss of smell. Patients with serious depression often complain to their therapists that they think their ability to detect odors is suddenly weakening. Are they losing their minds? No. Depression can truly bring about olfactory loss. Odor sensitivity tests on people who have been diagnosed with major depression show significantly diminished ability to detect scents at normal concentrations. The intimate and interdependent link between emotional and olfactory health is underscored even further by the fact that after treatment with antidepressant medication, smell sensitivity improves in these patients.[4]

Oddly, people who are plagued by a particular form of depression called *seasonal affective disorder* (SAD) show *increased* sensitivity to smells compared with their happy compatriots, and surprisingly this superior acuity exists year-round. This is surprising because SAD doesn't manifest like classical depression, which can happen at any time and usually lasts for many months or longer. Rather, SAD is a depression that only emerges in the winter, when light levels are low; in the springtime, people with SAD rebound to being normal happy people, or sometimes become manic—hence, the name *seasonal* affective disorder. The winter symptoms of SAD include: depressed mood, increased appetite, increased sleeping, and lack of motivation. Although SAD has prominent effects on mood, recent research suggests that it may primarily be a disorder of circadian rhythms, a suite of physiological rhythms that include when we feel sleepy and awake and that are highly influenced by light and fluctuate around a twenty-four-hour clock. When light levels are low, as they are in the winter, depression sets in, and when the hours of daylight increase in the spring, so does mood. The crucial role of light in SAD is further emphasized by the fact that people with SAD can be effectively treated with light therapy.* It also may explain why the sense of smell is not diminished as it is in classical depression. My belief is that rather than amygdala interactions, which influence the relationship between smell and classical depression, in SAD another limbic structure is at the crux of the depression-smell relationship—the *hypothalamus*.

You may have already noted that some of the depressive symptoms of SAD are different from those seen in major depression; for example, increased versus decreased sleeping and eating. The hypothalamus has primary control over our drives for eating, sleeping, aggression, and sex. In animals who hibernate, it is also the area of the brain that controls hibernation. Several researchers have suggested that SAD may be a psychological echo of hibernation. In keeping with this idea it turns out that irregularities in the functioning of the hypothalamus occur among people with SAD. Like all limbic structures, the hypothalamus interacts with the olfactory system. I believe that the hypothalamic aberrations in SAD may make the olfactory cortex more sensitive to smells than it normally would be, rather than less responsive as it is when the amygdala is malfunctioning.

Another biological process controlled by the hypothalamus is the menstrual cycle. Women's sensitivity to smells also varies with menstrual cycle phase, becoming particularly heightened around the time of ovulation when they are most fertile. This has

---

*Recent research suggests that SAD is most effectively treated with 10,000 lux for half an hour each morning, with the patient sitting about sixteen inches away from the light source (see Dr. Robert Levitan, Center for Addiction and Mental Health, University of Toronto).

significant biological implications because it turns out that smell is a critical factor in heterosexual attraction and finding the "right" mate, particularly for women, as you shall see. Women are also much more likely to develop SAD than men, and this isn't simply a corollary of higher head counts of depressed women than men. The same hypothalamic mechanisms involved in smell sensitivity and women's sexual physiology may also be responsible for altering smell sensitivity in SAD.

MORE THAN ANY OTHER SENSORY EXPERIENCE, fragrances have the ability to trigger our emotions: to fill us with joy and rage, to bring us to tears and make our hearts ache, to incite us with terror, and to titillate our desires. Have you ever been stricken with a feeling of dread, not known why, and then noticed a strange smell in the air? Thousands of New Yorkers had this exact experience walking in the streets and riding through the subway stops near the World Trade Center during the months after September 11, 2001. The strange charred and dusty scent was an instant reminder of that historical terror. On a more pleasant note, have you ever experienced the wonderful feelings of comfort and serenity that the scent of fresh, damp earth and moss invites after the rain? These examples illustrate how every day, scents affect our emotional lives in exceptional ways, triggering moods and emotional memories.

Not only do odors trigger emotions, they can also *become* emotions. My studies have shown that odors can literally be transformed into emotions through association and then act as proxies for emotions themselves, influencing how we feel, how we think, and how we act; I call this *odor-emotional conditioning*. In my laboratory we found that by pairing the feelings of frustration with an unfamiliar odor we were later able to make that odor alter behavior in accord with being in a frustrated mood. In one study, children who experienced an odor that had formerly been associated with trying to complete a frustrating maze showed less motivation and did more poorly on a simple test when exposed to that odor than children who had suffered through the frustrating maze but were exposed to a different odor or no odor during the test. In another study, adults exposed to a "frustration-associated odor" spent less time and were less motivated to solve challenging word problems compared with others who underwent the same procedures but were not exposed to this scent during the word problem test. For the disadvantaged participants, the scent had become conditioned to, and hence equated with, feelings of frustration, such that later the presence of the odor alone could influence the individuals to behave in a frustrated and unmotivated way. The odor caused the same behavior as experiencing the actual emotion would have.

Positive associations to odors can also lead to positive emotional conditioning, and there are many potential applications and benefits to society that could be developed from this connection. By linking feelings of intellectual competence to a specific odor and then using this odor when confronted with challenges at work or school, odor-emotional conditioning could be used to improve performance and productivity among individuals with low scholastic or job morale. Odor-emotional conditioning could also be used to improve social behavior in stressful settings. For example, by linking positive emotions to a specific odor, the odor could then be used to reduce violent or antisocial behaviors. In an effort to determine the effect of fragrance on social behavior, Susan Schiffman, a psychologist at Duke University, sprayed pleasant scents, such as chocolate chip cookies, into New York City subway cars and then tested whether the presence of these ambient aromas would make riders less aggressive. And indeed she observed that riders in the scented cars pushed, shoved, and made rude comments with almost half the frequency as riders in the unscented cars.

SCENT-EMOTION TRANSLATION

Why is it that our experience of scents and emotion are so interconnected, and what does this reveal about us and these two systems? I have a theory, based on evolutionary principles and insights into my own work, clinical findings, and research on emotion, that may explain why. A *chemical sense* was the first sense to appear in the mobile life-forms that emerged on the earth, and it is the only sense that the most primitive single-celled creatures share with us today. Its fundamental purpose was and is to detect chemicals to enable the organism to know what is good and what is bad "out there" for the basic goal of survival. Is this a good chemical (like food), or is this a bad chemical (like poison)? Do I approach or do I avoid? This is olfaction in its most basic form. From this very simple survival guide, the sense of smell has evolved to be a highly intricate go/no-go system for finding food and mates, establishing social hierarchies, assessing whether or not to be aggressive or fearful, avoiding predators, and many other complex behaviors. The sense of smell is the primary sense by which most of our animal brethren negotiate the world, including our primate relatives, and it is the sense to which they owe their survival. For us, vision has taken over as the primary sense that facilitates our survival. Yet odors still evoke in us the remnants of this primeval survival code.

The most immediate reaction we have to a scent is an assessment of *good* or *bad*. Approach what smells good, avoid what smells bad. Emotions also convey a simple message that is similar. Positive emotions such as joy and interest tell us to approach, to go forth and multiply—and ultimately result in our successful procreation and survival. Negative emotions such as anger, fear, and disgust tell us to avoid—and facilitate our survival by triggering a flight-or-fight response. Our emotions impart the same approach and avoid codes that smell imparts to other animals.

The connection between smell and emotion is not only metaphorical but also is founded on the evolution of our brain. A primitive olfactory cortex was the first fabric of our brain and from this neural tissue grew the amygdala, where emotion is processed, and the parts of the brain that are responsible for basic memory and motivation—the collective structures of the limbic system. In other words, the ability to experience and express emotion grew directly out of our brain's ability to process smell. I have often wondered whether we would have emotions if we did not have a sense of smell; *I smell therefore I feel?*

In my opinion, the human emotional system is a highly evolved, abstract cognitive version of the basic behavioral motivations instigated by the olfactory system in animals. Emotions are to us what scents are to our animal cousins. Smell for animals informs survival in direct and explicit ways; for us its primary survival codes have been transformed, into our experience of emotions. I call this *olfactory-emotion translation.*

Olfactory-emotion translation proposes that the sense of smell and emotional experience are fundamentally interconnected, bidirectionally communicative and functionally the same. I believe that the human brain has co-opted the survival-guiding olfactory system of animals into our survival-guiding system of emotions. Why do I believe this?

As I indicated earlier, we have discovered from patient research that with smell loss comes depression and with depression comes smell loss. With anosmia the olfactory neurons that normally also excite the amygdala are no longer activated, and over time this cessation in activity from the olfactory cortex causes atrophy and/or dysfunction in the amygdala—sowing the seeds for clinical depression. When depression comes first, I surmise that because the amygdala is functioning abnormally, the healthy activity that would otherwise be stimulating the olfactory cortex is skewed or compromised. That is, dysfunction in the amygdala alters the normal functioning of the olfactory cortex, and dysfunction in the olfactory cortex alters normal functioning of the amygdala. The olfactory and emotion areas of our brain are dependent upon each other for their mutual health and integrity. When one side malfunctions, it affects the functioning of the other. There is ultimate synergy between smell and emotion, but it is not always positive.

From a psychological perspective, our immediate responses to scents are simple, almost instantaneous assessments of liking or disliking; and many emotion theorists consider this fundamental "liking" response to be the basis of our highly complex emotional system. Moreover, the experiences of emotion and olfaction are similarly primal, visceral, and removed from verbal-semantic analysis. We have the same kind of difficulty using words to deconstruct our emotional experiences as we have articulating our experiences of smell. Try to explain why you love her, and what the attic smelled like. There is a fundamental analogy and bidirectional interaction between our experiences of smell and our experience of emotion.

The reason the bicycle accident that made Michael Hutchence anosmic was a crucial factor leading up to his suicide is because loss of the sense of smell brings with it severe disruption of mental health and happiness, while smell's intact state brings texture, richness, and a brilliant emotional quality to life in innumerable ways. The emotional enrichment imparted by scent is particularly striking for someone whose life is filled with its pleasures and vividness. Awareness and appreciation of scent actually increases our capability to smell. Recent studies at the University of California, Berkeley, have shown that when one is paying attention to and perceiving a scent, a part of the brain is activated that is otherwise quiet when that same scent is there, but you don't *notice* it.[5]

From biographical accounts it seems that Michael Hutchence was not simply aware of scents, he relied on them to intoxicate and transport him. Like Jessica Ross, life for Michael Hutchence without a sense of smell had become dark and dull, and the darkness grew larger and ever more enshrouding with the passage of time. For someone who took pleasure so seriously and could no longer indulge in one of its primary vehicles, it was like being a drug addict and losing the high from heroin—never able to regain that high and constantly seeking it, I believe desperation finally got the better of him.*

THE SMELL OF FEAR

The paradox for my proposition of a two-lane highway between smell and emotion is whether we can smell emotions. Are you afraid of dogs? If so, have you noticed that when you are around them, a so-called friendly pet has an uncanny tendency to single you out for torment? Or perhaps after being cajoled into horseback riding against your wishes, your horse, of all the ones in the ring, rears and tries to jump the fence. If these examples strike a chord, then you know that many animals seem to be able to detect our anxieties. What is it that these animals are sensing when they exploit our vulnerabilities? The answer is, they are smelling our fear.

The sweat on your tracksuit after a three-mile run is mainly composed of water, but when we are nervously waiting for medical test results, the sweat under our arms, and elsewhere, reflects the secretion of glands regulated by our nervous system and hence our emotional states. This sweat is more pungent than the sweat of exercise and has a characteristic odor, and like dogs and horses, we can recognize it, too.

---

*I do not claim that anosmia was the sole reason for Michael Hutchence's suicide, but that it contributed to a depressive mind-set that made it more difficult for him to cope with a number of negative factors that were impinging on his life at the time.

Research by Denise Chen, at Rice University in Texas, has demonstrated that humans understand the scent of fear, as well as happiness, in one another. To investigate whether we can smell each other's emotions, she had college students watch excerpts from happy and frightening movies while they wore gauze pads in their armpits. The pads were then collected and aggregated by gender for happy and fearful male and female sweat. Another group of young men and women were then asked to use their noses to judge which emotional category the underarm pads were from. Women were able to judge happy sweat from both male and female donor pads, while men were better at identifying the scent of happy rather than fearful women. But both men and women were best at recognizing the scent of male fear. Female fear sweat was also recognizable but because male sweat is usually stronger than female sweat, it was easier to classify.[6]

THE HARDWARE

What are these smells that can comfort us, frighten us, and inspire our deepest passions? Smells are chemicals, and they can be extremely complex and contain thousands of molecules, like the rose scent emanating from your flower bed, which is made up of between twelve hundred and fifteen hundred different molecules. Or they can be very simple and comprise just one molecule, like phenylethyl alcohol, the chemical that imparts the scent of rose in many commercial hand lotions. Surprisingly, my research has shown that we cannot reliably tell these two versions of an odor apart—the synthetic fake rose and the rich, natural flower bed—and if asked which is the fake and which the real thing, we are more likely to err in favor of exclaiming the artificial aroma as being "the real thing." This is because we are often more familiar with the manufactured version of "natural" scents than the natural scents themselves and so these synthetic copies become the prototypes for what we believe these fragrances *should* smell like.

Not all chemicals, whether a single molecule or massive bundles, can be smelled. For humans to be able to smell a chemical, it must be of low molecular weight,* volatile, and able to repel water, so that it can stick to our olfactory receptors. However, there are also exceptions to the rule, and we don't detect all small, airborne, greasy chemicals. One common example is methane (natural gas). We also can't smell carbon monoxide, which is a by-product of methane. If you use natural gas in your home, you may think that it has a skunky/rotten-egg smell, but this is only because utility companies add a compound, tertiary-butyl mercaptan, to natural gas so that you can "smell" if your pilot light has gone out. In addition to not being able to smell carbon monoxide, we also cannot smell the gases that make up the air we breathe. Pure air is made up of about 78 percent nitrogen, 21 percent oxygen, and 1 percent a mixture of eight other gases including carbon dioxide, helium, and neon, none of which we can smell.

---

*Molecules that have a molecular weight of greater than 350 daltons cannot be perceived through our sense of smell. One dalton is equivalent to the weight of a hydrogen atom (1.657×10^{-24} grams). A simple water molecule has a molecular weight of 18.

Every day we inhale at least twenty-three thousand times, and with each breath comes the opportunity to perceive the chemicals whispered by a rose on a soft summer night, the chemicals nestled in the crook of your lover's neck, the cake baking in the oven, or the effluents of dirt, mold, gas, and decay—every scent that you have ever smelled. With each breath, air containing odor molecules enters our nostrils and is swept upward into the nasal passages of the nose (see page 21). In addition to being the location of the olfactory receptors, the function of the nose is to filter, warm, and humidify the air that we breathe. The interior of the nose is not simply a smooth cavity but possesses a complex geometry of ridges shaped by the underlying turbinate bones. The geometry of our nose, and the air convection system it generates, is more complex than the airflow around an aircraft wing. This special architecture allows the air containing odor molecules to pass through a narrow space called the olfactory cleft, where the chemicals then make contact with the olfactory receptors.

The olfactory receptors are buried in two patches of yellowish mucous membrane, called the *olfactory epithelium,* about seven centimeters up from each nostril. The olfactory receptors are on the tips of the olfactory neuron dendrites. We have approximately 20 million olfactory receptors covering the epithelium of both our right and left nostrils. However, our 20 million receptors would look paltry to a dog, like the bloodhound, who has about 220 million olfactory receptors. Olfactory receptors, unlike the receptors in any other sensory system, are directly exposed to the outside world, which is why among other things we can inhale drugs.

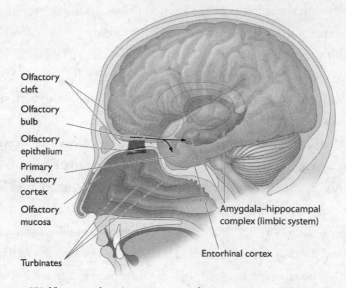

Olfactory
cleft

Olfactory
bulb

Olfactory
epithelium

Primary
olfactory
cortex

Olfactory
mucosa

Amygdala–hippocampal
complex (limbic system)

Turbinates

Entorhinal cortex

From Wolfe et al.: *Sensation and Perception*, 2006; Sinauer Associates, Inc.

After the chemicals that we smell make contact with the receptors on the olfactory epithelium, nerve impulses are passed from the olfactory receptor axons through the cribriform plate and into the olfactory bulbs. From the olfactory bulbs, scent information is then relayed to the olfactory cortex and the limbic system, and from there on to diverse regions of the brain, including the vision, taste, and touch centers—which is why, for example, we find that food artfully arranged on a plate "tastes" better than if it is all mashed up.

We have more receptors for smell than we have for any other sense except vision, yet the area of our brain that is devoted to olfaction is very small, about 0.1 percent of our total brain size.

The fact that such a small proportion of our brain is devoted to the sense of smell has led to the myth that we are not very good at smelling, but this statement is only true in relative terms. Compared with dogs and mice our noses aren't especially keen, but we can still sniff out most of the scents that a bloodhound can; we just need a much higher concentration of the odor to do so. Dogs can detect odors at concentrations nearly 100 million times lower than we can. This is the equivalent of being able to detect a drop of chocolate in a city the size of Philadelphia. Salmon are also amazing smellers and can detect the scent of their birth waters at a concentration as low as 3×10^{-18}. That is, 1 milliliter of scent diluted in 333,333,333,333,333,333 milliliters of water. The average nonchalant human nose can discriminate between ten thousand and forty thousand different odors, and professional smellers—literally called "noses" in the fragrance industry—and perfumers, whisky blenders, and chefs may be able to discriminate upward of one hundred thousand odors. The only real difference between an expert and an average nose is training. Technically, anyone with a healthy olfactory system can perceive an enormous number of aromas, the only limiting factor being the number of smellable molecules that are out there.

FROM THE BAKERY DOOR TO CROISSANT CRAVING

The basic anatomy and physiology of our sense of smell is well understood, but oddly the deeper question of how a set of chemicals goes from wafting out of a bakery door to calling out "croissant" to you still remains a mystery. For all our other senses we have a complete understanding of both physiological and psychological processes, but in olfaction we are lagging far behind.

The reason we are delinquent in understanding our most basic sense is that the sense of smell was not deemed worthy of study until very recently. The concept of "odor" was branded with negative connotations in the Victorian era. Then the notion that animals smell—that is, stink—but *civilized* people did not, or rather "should not," reigned, and this view has contaminated endeavors to explore and understand our sense of smell ever since. Research psychologists and neuroscientists are also to blame for why olfaction has been relegated to the sensory dustbin. The long-standing excuse has been that odors cannot be properly studied because their physical properties are very hard to precisely control and the responses to them are subjective; therefore, studying smell is "unscientific." It is true that to impose the same level of physical rigor in testing olfaction as one could when testing hearing, for example, is difficult, but that doesn't mean that the scientific investigation of olfaction is inherently compromised. It just means that you have to ask somewhat different kinds of questions, and use somewhat different techniques, as you will discover throughout this book. And the appropriate questions and methods to study olfaction aren't restricted to psychological science; they include fundamental biology and chemistry as well.

The decades of waving away the scientific legitimacy of olfaction ended with a breakthrough discovery in 1991 that not only made olfaction "scientifically" scrutinizable but made it one of the hottest and most desirable areas of research to be in. Linda Buck and Richard Axel, working at Columbia University,

versity, published a paper in the journal *Cell* that led to their winning the Nobel Prize for Physiology or Medicine in 2004.[7] It was the first time ever that a Nobel Prize had been awarded to anyone or anything having to do with the sense of smell. Buck and Axel discovered that there are about a thousand different odor receptor types, each coded for by a different gene. What this means is that in the nose (their test animal was the mouse) there are one thousand different types of receptors that decode the chemicals in the air into smells and that each of these receptors is represented by a different gene in the body. For comparison, in vision there are only four types of receptors—rods for black-white (night) vision and three types of cones, sensitive to short, medium, and long wavelengths of light, for color vision.* Buck and Axel also discovered that there were a lot of *pseudogenes*—DNA sequences that are remnants of genes that are no longer functional—in that family of one thousand olfactory receptor genes. In a dog, about 15 percent of its olfactory receptor repertoire are pseudogenes; the mouse has about 20 percent. In humans, the number of pseudogenes is much higher—about 65 percent. Precisely how many olfactory genes are functional in humans is currently under debate but it is on the order of three hundred to four hundred. Even though between 60 and 70 percent of our olfactory genes don't work, 1 percent of all the genes in our body is committed to regulating olfaction, which is far more genetic devotion than anything else in our bodies and brains gets.

---

*Short wavelength corresponds to seeing violet-blue, medium to green-yellow, and long to orange-red. A new photosensitive ganglion receptor cell has just been identified in the retina that modulates circadian rhythms.

Until very recently it wasn't understood why we would have so many olfactory genes that no longer do anything. But in January 2004, researchers at the Weisman Institute in Israel proposed an answer. Yoav Gilad and his colleagues observed that in Old World primate species such as gorillas and rhesus monkeys about 30 percent of their olfactory receptor genes were pseudogenes, whereas most New World species (e.g., squirrel monkeys) have a lower proportion—around 18 percent. The one New World exception is the howler monkey, which has around 30 percent of its olfactory receptor repertoire as pseudogenes. It turns out that the howler monkey and Old World primates have something in common with us—trichromatic (red, green, blue) color vision. Other mammals who lack trichromatic color vision, such as mice and dogs,* also have few olfactory receptor pseudogenes. The theory is that with the emergence of full color vision we lost the need for detecting the world so keenly with our sense of smell, and there was essentially an exchange in importance between these two senses in primate evolution. The better you can see, the less acutely you need to smell. Animals, including humans, either have excellent color vision or an excellent sense of smell, but not both. The finite size of the human brain is to blame. The human brain is limited to between 1,300 and 1,400 grams in weight. Having a highly complex sense of both smell and vision would take up too much brain space and so these functions had to compete with each other for which had the better survival value. It seems superior visual detection was better for our ancestors' survival than superior olfactory acuity, and the advantages offset the

---

*Dogs are red-green color-blind.

limitation to our sense of smell; hence, the large proportion of pseudogenes in our olfactory code.

Buck and Axel's seminal discovery of olfactory receptor genes produced a research explosion in the molecular biology and biochemistry of smell. With this energetic work we are now closer than ever to understanding how olfaction works, but a final theory for how molecules are translated into perceptible scents has still not been fully realized.

The most established current theory is referred to by both its advocates and detractors as *shape theory*. According to shape theory, the key to odor translation is the match between the shape of odorant molecules and odor receptors. This idea is strikingly similar to an ancient theory of smell involving shape that was suggested by the Roman philosopher Lucretius. In its modern form, shape theory got its legs in the 1960s from John Amoore, an independently minded biochemist, olfactory theorist, and entrepreneur. It was then further developed and refined by Gordon Shepherd and his colleagues at Yale University during the 1970s. In a nutshell, modern shape theory contends that odor molecules have different shapes, and odor receptors have different shapes; the ability for an odor to be detected is determined by how a specific molecule is recognized by specific olfactory receptors. From the most recent molecular research it seems that the detection of a scent is done by a combinatorial code. Different scents activate different arrays of olfactory receptors in the olfactory epithelia, producing specific firing patterns of neurons in the olfactory bulb. The specific pattern of electrical activity in the olfactory bulb then determines the particular scent we perceive. The scent of a mango elicits a different pattern of neural impulses than the smell of skunk.

One of the surprising results to emerge from the recent receptor work is that changing the concentration of an odor will also change the receptor code. That is, a weak concentration of baking butter sets off a different firing array than a high concentration, which explains why you need to be within a certain proximity of the bakery door to have the eureka moment of "croissant." The link between receptor code activity and intensity may also explain the differences in olfactory prowess between dogs and us. A dog is not necessarily detecting molecules that we cannot physically smell; rather, he is detecting them at a very much lower concentration than we are and therefore is capable of smelling things that are just "not there" to us. If all of our one thousand genes coded for functional olfactory receptors, our scent sensitivity would be completely overwhelming and all consuming, and I am sure human culture, civilization, and our experience of reality would be very different from what it is now.

In spite of growing support for shape theory, it is not accepted by all as the answer to how chemicals are detected by the cells in our noses and then become recognized by the brain as a croissant or kiwi. An alternate theory that has raised its head several times throughout history is that our perception of smells is based on the different vibrational frequencies of the molecules that compose various odors. In recent years this idea has been championed by Luca Turin, the emperor in Chandler Burr's *The Emperor of Scent*. In essence, *vibration theory* proposes that due to their atomic structure, there is a different vibrational frequency for every perceived odor molecule, and molecules that produce the same vibrational frequencies will have the same smell. Turin reported that chemicals that have predictably similar vibrations due to their molecular composition also have similar smells. For example, all citrus-smelling odors are in the same class of vibrational frequency. But independent researchers testing this theory have not validated his claims. There has been much less research pursuing *vibration* than there has been for *shape,* so current research trends may unfairly bias our understanding. Nevertheless, vibration theory, unlike shape theory, cannot explain several conundrums of olfactory perception, such as "specific anosmias" and the different scents produced by stereoisomers.

Stereoisomers are molecules that are mirror-image rotations of one another, and although they contain all the same atoms, they can smell completely different. For example, *d-carvone* (the right-handed isomer) smells like spearmint, and *l-carvone* (the left-handed isomer) smells like caraway. Shape theory explains this as being because the rotated molecules do not fit the same receptors—like trying to put your right hand into your left-hand glove—and thus different receptors are activated for these two molecules and different scents are perceived. Vibration theory cannot explain why stereoisomers smell different from one another because the vibration of stereoisomers is the same.

Another mark in favor of shape theory comes from the study of specific anosmias. A *specific anosmia* is the inability to smell one specific compound with otherwise normal smell perception. Most specific anosmias, which appear to be genetic, are to steroidal musk compounds—animalic-sweaty scents within a specific chemical family. The most studied specific anosmia is an inability to smell the steroidal musk *androstenone.* Fifty percent of the population has a specific anosmia to androstenone. However, of the remaining 50 percent, about half describe

the smell as a *sweet musky-floral,* while the other half describe
it as an unpleasant *urinous* odor. Vibration theory cannot explain why some people perceive the scent as a sweet-floral, others as urinous, and why yet others can't smell it at all. However,
shape theory can account for these observations by proposing
that in individuals with this specific anosmia the receptors that
detect androstenone are nonfunctional, while among those who
can smell it, different receptors are activated in those who perceive floral than in those who perceive urine.

From where we are now, shape theory has better support
and explanatory value, but it is also true that all molecules vibrate, and all receptors are made of atoms that vibrate as well.
This leaves room for vibrational interactions between odor molecules and receptors to play a role.

Beyond the unresolved theory for the translation between a
chemical in our nose and a smell percept in our olfactory bulbs,
the more profound question of how the pattern of activity in our
olfactory bulbs becomes psychologically organized, processed, interpreted, and recognized by you as "croissant" with all its motivational and hedonic sequelae remains unknown. What we think a
certain scent is, its connection to language and concept, what the
scent means to us, what it makes us feel, and what it reminds us
of—all interact in a complex multifaceted dance and determine
our *perception* of that scent. Odor sensation happens at the level
of our nose and olfactory bulb, but olfactory perception occurs in
our mind, where our personal experiences with scents take over.

CHAPTER 2

AS YOU LIKE IT

A rose by any other name would not smell as sweet.

—*(not)* SHAKESPEARE

When I was seven years old we moved to Montreal. The move was difficult for me because it rode on the heels of several international relocations, during which time I had acquired expert knowledge of the cruelty of children toward the outsider. Our move to Montreal was to be permanent and I longed to integrate into the schoolyard cliques, but the more apparent I made my wish for friendship, the more rejection was heaped upon me. Suffice it to say that I spent my first year in Montreal feeling lonely and frustrated. One afternoon, during those first few months, a salesman rang our doorbell—he was selling hair products from some new company. I remember being startled that my mother actually purchased this particular shampoo and conditioner, as it was unlike her to spend money spontaneously on door-to-door sales. And I remember that the word *aquamarine* was in the product name. The rest of the details are a blur, but that purchase changed my life, and the ensuing experience became indelibly imprinted in my memory.

The night of the purchase I asked to try the new hair products. I was captivated by the shiny exotic bottles and eager to experiment, to experience something pleasurable and atypical in my day. I remember squeezing the translucent turquoise shampoo into my hand and lathering it into a bubbly and delicious foam in my hair, but most of all what I remember is the exquisite scent that arose from the bubbles. I had never smelled anything like it before—sweet, piney, watery, and mysterious—and it seemed to me to be the most sublime aroma on Earth. While drenching myself in my parents' pink tile and chrome shower, which I was allowed to use only on special occasions, for that first hair-washing experience with this superb new smell, I became intensely happy for the first time in a long time. I don't think my parents were particularly impressed by these hair products since they stayed in their shower diminishing very gradually for almost a year. While they lasted, I would periodically ask to use them so that I could relive that wonderfully transporting fragrance experience. I also began to realize that if I wanted to feel better after a bad day in the schoolyard, all I needed to do was open the lid of one of the bottles and inhale its sublime scent. After the last fragrant drops were spent, I would pester my mother now and then to buy those special hair products again, but no one ever returned to our house to sell them, and they were not available in any stores. That one bottle of shampoo and conditioner were all I ever knew, and though I have tried many times over the intervening years to find them, I have never had any luck. Nevertheless, ever since that first experience with "aquamarine," if I smell anything that has any resonance with that aroma, I instantly love it and will go out of my way to buy it or find a way to use it.

As I confessed in the preface to this book, another scent that I truly enjoy is the smell of skunk. Although I wouldn't choose a shampoo with this aroma, when I am out in the country and the scent of this animal is carried on the breeze, I find it very pleasurable. This is more unusual than liking a shampoo fragrance but there are many others who, when feeling safe enough, admit to me that they share this unusual preference. And it isn't that I like all smells. I find creosote—that smell of asphalt being laid down—extremely unpleasant.

Why am I so fond of certain specific aromas and dislike others? Why are you? Are we born liking and disliking various smells, or do our aroma preferences come about in another way? The general assumption, though never scientifically proven, is that we have an innate, hardwired predisposition to like or dislike various odors. This presumption is largely based on the fact that our responses to tastes, particularly those to sweet and bitter, are innately positive and negative to us, respectively, and preferences for scents have been presumed to follow suit. This popular notion is further exacerbated by the incorrectly blurred distinction between the senses of smell and taste.

I disagree with the view that we come into this world with a set of odor likes and dislikes—that the scent of rose is good and skunk, bad—and, rather, I am convinced that our aroma preferences are all *learned*. Now let me convince you.

If one wanted to prove that the liking or disliking of odors was innate, the best place to start would be with infants, because if anyone is going to show spontaneous innate reactions, it would have to be newborns, since they haven't learned much yet. Several researchers have examined newborn odor preferences and have found that infants' responses to smells do not match the responses of adults to those same scents. For example, infants like the smell of feces and are equally indifferent to scents that adults view as negative or positive, respectively, such as rancid cheese and banana. In fact, it is not until children are about eight years old that they start to show odor preferences that match the responses of the adults in their culture. It turns out that there is *no* data that infants show predictable—innate—reactions to smells. But there is considerable evidence that what infants and children like to smell is due to their experiences; in other words, learning, and this learning begins even before birth.

Our olfactory system is the first sense to develop. In fact, we have a fully functioning sense of smell by the time we are twelve weeks in the womb. This is in stark contrast to our visual system, which takes several years after we are born to become fully mature. The fact that we have a completely functioning sense of smell when we are just three months postconception means that we can begin to learn about odors well before we are born; and we do. The way we come to know various aromatic chemicals during this prebirth period is through what our mothers consume. Amniotic fluid, which surrounds the developing fetus, is composed of what is in the mother's body as well as the fetus's body. So what a mother ingests becomes chemically present in her amniotic fluid, and whatever aroma molecules are there can be perceived by her developing child.

Several studies have shown that what a mother consumes while she is pregnant will influence her baby's flavor* preferences after he or she is born. Julie Mennella, a longtime researcher on the development of flavor preferences, and her colleagues at the Monell Chemical Senses Center in Philadelphia found that mothers who ate garlic, drank alcohol, or smoked cigarettes while they were pregnant had infants who preferred these scents compared with infants who had not been exposed to these smells during their gestation. The infants' preferences for these scents even superseded their liking for vanilla, which is often believed to be an ubiquitously favored aroma. Breast-feeding can similarly influence the development of flavor preferences, including for healthy foods. Julie found that mothers who drank carrot juice while breast-feeding had infants who liked carrot flavor when later exposed to it as "baby food" much more than infants who did not have this early exposure.[1] The implication is that one could have a child who likes liver and spinach if the mother eats liver and spinach during these early moments. It also so happens that these early learned flavor preferences can carry on into adulthood and have subtle long-lasting effects on our culinary orientations.

E. P. Köster, a smell researcher in the Netherlands, knew that children in Germany who were fed with formula experienced a vanilla-flavored drink—and that this product had been used for

---

*Flavor is a combination of basic tastants (salt, sour, sweet, bitter) plus smell, and it predominantly relies on smell. See Chapter 7 for more details.

several decades. To see how this early experience with the flavor of vanilla might have impacted later flavor preferences, he conducted a clever study at a fair in Frankfort, Germany. In the late 1990s, Köster and his colleagues set up a booth of ketchup samples. One cup contained regular ketchup and another cup contained ketchup flavored with vanilla. One hundred and thirty-three men and women in their late twenties and early thirties came to sample the two versions of ketchup. From a questionnaire distributed to the fair attendees when they first entered the grounds, Köster's group knew who had been bottle-fed and who had been breast-fed as infants. Köster found that those who were bottle-fed had a striking preference for the vanilla-flavored ketchup compared with those who had been breast-fed.[2] A new idea for Heinz? Endless commercial possibilities could be exploited if flavor companies knew all our earliest flavor experiences.

PERHAPS YOU ACCEPT MY ARGUMENT that children's responses to odors are learned and variable, but still question whether adult responses to "very bad" odors could really be due to learning; *how could anyone find the smell of rotting bodies okay?* But adults across this planet do not agree on what smells good and what smells bad—even for the stench of death. Culture, which conveys another form of learning, explains how and why.

The adage that *one man's meat is another man's poison* has its origins in astute observations of reality. Asians consider the smell of cheese to be hideous, yet Westerners regard it as anything from comfort food to a sumptuous indulgence. In contrast, the Japanese enjoy a meal for breakfast called "natto," a fermented soybean dish, which this North American couldn't bear to bring near her mouth. Natto smells like burning rubber to me, and although burning rubber may not be an unpleasant scent to some, it is not an aroma that I connect in any way with eating. The fruit durian (also called jackfruit), which is found throughout Southeast Asia, is a local delicacy there, but when Westerners smell it they are repelled. In *A Natural History of the Senses,* Diane Ackerman described the scent of durian as a cross between "a sewer and a grave"[3]—a vividly ghastly combination—and yet thousands of people consume it daily and *like* it. If you're thinking that the example odors I have been mentioning really aren't *that* bad, and that there must be consensus on truly horrid stenches, this also doesn't seem to be the case. Fecal smells are not high on most North Americans' best smells list, but the Masai like to dress their hair with cow dung as a cosmetic color treatment, and to the U.S. military's great surprise, it has been impossible to develop a universal "stink bomb" to use for crowd dispersal, as a safer alternate to tear gas. In a recent study undertaken by the U.S. military, no odor tested, including "U.S. Army issue latrine scent," was found to be unanimously unpleasant across people from a range of ethnic groups.*

---

*The most recent research on stink bomb creation by Pamela Dalton of the Monell Center suggests that if you mix a cocktail of "noxious" and unfamiliar odors together, you are more likely to get unanimous dislike than if you use only one scent. The reasoning is that (1) it is harder to acclimatize to a mixture than to a single scent, and (2) if you have at least one odor in the mixture that is unfamiliar to at least one group of people, the mixture will probably also smell unfamiliar and hence be more likely to be considered objectionable. The key here is that what is *unfamiliar* will be more readily disliked than liked.

The scent of carrion has also been touted as one to which there *must* be innate abhorrence, but again we find that culture, in this case defined historically, refutes the presumption. In Europe, before the invention of refrigeration, rotted meat and fish were often sold. Although many recipes of the time offered suggestions for disguising the taste of spoiled meat, perhaps because of its prevalence or out of necessity, historical accounts indicate that many people actually *preferred* their meat putrid so that when it was served it gave off a strong, "high" aroma and therefore signified a richer and more robust meal.

So why is it that I like the smell of skunk while an Asian will categorically refuse Roquefort and others find the aroma of rotting meat and human waste acceptable? And how is emotion involved? A theory first formulated by Trygg Engen, the father of psychological research on smell, proposes that before you have experienced an odor it is inherently meaningless, a *tabula rasa*, a blank slate; however, once you experience it, the context (place, situation, person, or event) in which you perceive it and, most important, the emotional value of that context become attached to that aroma, and henceforth the odor takes on that emotional significance and meaning and becomes liked or disliked accordingly. The smell of coffee is stimulating and pleasant because the physical and emotional effects of drinking coffee have become attached to its aroma. This theory for how we acquire odor hedonic responses is called *odor-associative learning*.

Here is a classic example of odor-associative learning in action. Suppose that you have to undergo an unpleasant, anxiety-producing event, such as a surgical procedure in a hospital.

This is the first time that you have been to a hospital, apart from your birth, and when you enter it you notice its distinctive and *novel* aroma. First encounters, that is, novelty, are especially important in the development of odor preferences because if a smell is already familiar to you, you would already have associations with it. The issue of novelty is also why most of our odor preferences seem to come from childhood, because this is a time when most of our sensory experiences are new. Anytime in life that we encounter a new smell, emotional odor-associative learning can take place to determine our hedonic responses to it.

Forty years ago in the United States, "hospital odor" was typically a combination of standard-issue disinfectants, medicines, body odors, and ether. "Hospital smell" does not have to be the same concoction across time and place; the only requirement is that the aroma be unfamiliar before your visit to the hospital. The point is that although this scent is not inherently good or bad, the context you are in when you first consciously experience it is bad, and therefore this initially meaningless odor acquires the negative valence of your feelings. "Hospital scent" becomes associated to hospital feelings and through this emotional connection acquires its emotional and hedonic significance—*unpleasant*. Whenever you smell this scent again, your feelings of hospital aversion resurface, and since the location you are most likely to smell it in is a hospital where negative things are likely happening, this unpleasant odor association is further reinforced.

Hospital smell, and the general loathing of hospitals it brings, is well known to hospital administrators; as a means to counteract this, a tactic currently being used by some hospitals is to utilize cleaning and deodorizing scents that are already familiar and "pleasant" to us. This is intentionally done to eliminate the formation of bad feelings attached to hospital smell. What is clever about this idea is that it is quite difficult to form a new association to a scent once it has been learned and first associations have already been made. Olfaction is unique among the senses in the strength with which first associations interfere with forming new ones. In practice, therefore, the familiar scent of vanilla would be difficult to pair with and then elicit "surgery worry."

The central tenet of odor-associative learning is that how you *feel* when you first encounter a particular scent determines your future hedonic perception of it. Odors that we like are ones that we first encountered in a situation where we were happy or are connected to something with positive meaning, and odors that we dislike were first encountered when we were in a negative emotional state or are connected to something with unpleasant meaning. Perhaps the best example to illustrate how emotional learning is behind our aroma preferences is the story told to me by a young woman who explained that she hated the smell of roses because the first time she smelled them was at her mother's funeral.

We can use the infants who like garlic to dissect this process in detail. These babies had mothers who ate a lot of garlic during their pregnancy and/or while they were breast-feeding, and thus their infants' first encounters with this scent were with Mama and with food. Eating has critical motivational value and so does Mama herself, signifying love, protection, affection, and nourishment. Therefore, when Mama is paired with a scent, this scent becomes positive. These early maternal experiences also underlie why vanilla is typically touted as a scent with universal appeal. To the extent that this is true it can be explained by the fact that vanilla is a volatile flavor component of human breast milk as well as many milk formulas.

The fact that maternal diet can influence a newborn's flavor preferences also means that cultural differences in response to foods and aromas can set their roots prior to birth. A culture with a diet rich in spices will produce babies wanting curry with their mashed bananas, whereas a culture in which the diet is relatively spice free will produce babies who snub foods with seasonings.

So where do my responses to aquamarine, skunk, and creosote come from, and how do they fit the odor-associative learning explanation? The "special occasion-ness" of it all—being able to use my parents' deluxe shower and the pleasure of the new exotic shampoo that I had never experienced before—clearly initiated my love of aquamarine scent. Moreover, the escape from the gloom that I usually felt during those days made the happy feelings especially salient. Similarly, when I was in the car on that lovely summer afternoon and my mother exclaimed, "Oh, I love that smell," a link between my feelings for my mother, the bucolic context, and that specific aroma was formed. And so I love the smell of skunks. By contrast, creosote is unpleasant to me because of another car trip where I felt very queasy and hot and we were stuck in traffic while asphalt was being poured. The association between nausea and creosote scent instantly bonded. If the creosote and skunk car ride situations were reversed, I am sure I would love the smell of creosote and be nauseated by the scent of skunk.

CULTURE AND CONTEXT

Imagine the smell of burning leaves (my apologies to those readers who did not grow up knowing this scent). The smell of burning leaves is such a well-liked aroma to many that Christopher Brosius, the eccentric perfumer and founder of Demeter Fragrances, known for creating naturalistic and unusual perfumes, such as Dirt, concocted the fragrance Burning Leaves due to customer demand.* However, the real smell of burning leaves is actually full of noxious toxins and pollutants, which is why it is now illegal to burn them in most U.S. states. How can it be that so many people like such an unhealthy brew? The answer is in your past. Hearken back to your childhood and try to recall your first experiences with this scent to figure out why *you* like it. Perhaps it was connected to the fun of Halloween. Or it signaled the family festivity and food of Thanksgiving. Or maybe it was linked with a special time you shared with a parent or grandparent. Or it was simply the background aroma of carefree afternoons when the weather turned cool and beautiful colors stippled the trees. Have I struck a chord?

Since it is now illegal to burn leaves in most U.S. states, children growing up in America today are unlikely to encounter this scent, and if they do, they will probably know that burning leaves emits harmful chemicals—and therefore is "bad." If we

---

*Demeter's Burning Leaves is now called Bonfire.

fast-forward twenty years into the future and give young adults the smell of burning leaves to evaluate, would they like or dislike it? Would they even know what it was? Will there be any demand for Demeter's perfume of burning leaves in twenty years?

In addition to learning through our own experiences, culture provides a general umbrella under which a variety of our proclivities are acquired, and the differences between cultures don't need to be large for large differences in aroma preferences to be seen. Consider Britain and the United States. A supermarket poll reported by the *Times* of London (January 2004) revealed that in Britain the top ten favorite smells were: fresh bread, frying bacon, coffee, ironing, cut grass, babies, the sea, Christmas trees, perfume, and fish and chips. We may agree with some of this list, but I can guarantee that someone from Britain won't agree with an American top ten. Wintergreen mint is a favorite American scent, yet it is highly disliked in the United Kingdom.

Why don't the British like the smell of wintergreen candy, while Americans do? In the mid-1960s, a study was conducted in the United Kingdom where adult respondents were asked to provide pleasantness ratings to a battery of common odors.[4] A similar study was conducted in the United States in the late 1970s.[5] Included in both studies was the odor methyl salicylate, the scent of wintergreen. In the British study, wintergreen was given one of the lowest pleasantness ratings out of a sample of many odors, whereas in the American study it was given the highest rating of the whole sample set. How could it be that two cultures separated by a common language could show such big differences in odor preferences? The answer lies in history. In Britain, the scent of wintergreen is used in various medicaments, and particularly for the participants in the 1966 study was characteristic of rub-on analgesics that were popular during World War II. Conversely, in the United States, the smell of wintergreen is almost exclusively found in candy and gum, items with sweet, positive experiences attached to them. The British don't like wintergreen because it brings up feelings and associations connected to medicines and wartime, whereas North Americans like it because it conjures associations of sweetness and treats.

Cultural differences in experience also explain why a candidate aroma for the universal stink bomb has not been found. Imagine that you lived in a culture where modern sewage systems didn't exist; you would be very familiar with aromas such as "U.S. Army issue latrine scent," and they may signify nothing more than daily life—and be as ubiquitous as the smell of gasoline is for us. This "ubiquity" is in fact why Beijing lost the bid for the 2004 Olympics. In 1993, when the bid was made, only 30 percent of Chinese homes had private toilets. Instead, public toilets were the norm and unlike the restroom in your local park, these public lavatories were pits in the ground, in an open, walled-in room, with no individual stalls and no running water. A consequence of the omnipresent public lavatory is that the air that hung over Beijing was conspicuously heavy with toilet aroma; imagine what it would be like in the summer. After the bid loss and realization of the need to renovate the conditions of public lavatories, the Chinese government began spending taxpayers' money on installing modern toilets. Surprisingly, many people complained that their money would be much better spent elsewhere. In other words, the average Chinese citizen really didn't mind the stench, not to mention the lack of privacy. The public outcry was ignored, and by 2004 many public lavatories had been eliminated and 80 percent of homes boasted private facilities. The Chinese government has vowed that by 2008 no one in Beijing will be more than an eight-minute walk from a flush toilet.

How about an even more dreadful example—the smell of a burning body? This is a scent that most Westerners could never conceive of liking. But what if you live in a country where public cremations are the typical ritual for laying the dead to rest, such as in India? In this case the scent of a burning body would be familiar, and as indicated in the previous example, odor familiarity is highly correlated with odor acceptance. Moreover, many ritual cremations are accompanied by celebration—like being at an Irish wake—and if this were your experience, festive emotions would become attached to this smell, and hence you would have actually learned to like it.

By contrast, lack of personal history or cultural connotations offers a glimpse into one's "natural" response to an odor, and I recently had the opportunity to observe this with my favorite contentious aroma—skunk. A few summers ago some skunk-naïve Swedish colleagues were visiting me in Rhode Island and we walked down a street where a skunk had recently declared his presence. Although the degree of liking varied, none of my visitors said that they found the scent unpleasant. People from Europe, where skunks are not indigenous, typically do not find this odor to be bad unless previously warned.

You may be wondering how an animal whose natural defense against its predators is to repel them with its "bad" smell would not elicit this reaction from us. The reason is because the skunk's natural predators are hardwired to be repelled from this scent, but we are not one of them. What makes an animal have a hardwired scent response or not has to do with what kind of ecological habitat it calls home. We are "generalists," animals who can successfully exploit any habitat on Earth, and as such we have to *learn* what is good and what is bad. The scent of "poison mushroom" in one locale could be the scent of "nutritious food" in another. However, animals who are "specialists," who live in defined and specific ecological niches, are born knowing what scent to approach and what scent to avoid, as you will soon see. The skunk and its natural predators are specialists, and to aid the skunk's survival it evolved a mechanism to deter its would-be menacers—that aromatic spray. There is, however, a feature of skunk spray that will automatically repel us, too, but you have to be up close and personal. Skunk spray contains chemicals that activate the tactile pain system around our face. Skunk spray in your eyes will burn and encourage you to get away from it, but not because of the smell per se.

M Y LABORATORY HAS DIRECTLY TESTED the odor-associative learning theory. In a recent study, we paired several unfamiliar odors with either a good or bad emotional experience: playing a frustrating computer game and losing play money, or playing a fun computer game and winning real money. We then compared people's ratings of the odors before and after the emotional association. The computer game experiences were hardly intense compared to real-life travails;

nonetheless, we found that after being paired with the annoying computer game, the odor was disliked more than it had
been to start with, and after being paired with the winning
game, it was liked more.[6] In another study examining physiological responses to smells, researchers found that the scent
of eugenol, which is the "clove" odor found in dental cement,
was evaluated more negatively and elicited autonomic fear responses, such as sweating and rapid heartbeat, among people
who had bad experiences at the dentist, while unafraid, cavity-
free participants showed no such reactions.[7] For fearful patients the association of the dentist's drill had become connected
to the coincident smell such that eugenol itself could subsequently elicit fear reactions. For those participants who hadn't
had any painful fillings, and hence bad associations with eugenol, this scent was just a neutral, nonmeaningful aroma.

Still have your doubts? You may be thinking, *how is it that
I have opinions about odors that I have not had dramatic emotional encounters with, or for that matter, may have never even
smelled?* Learning can still explain these responses, in these
cases learning through social transmission and cultural norms.
What your culture tells you is good or bad becomes incorpo-
rated into your perception lexicon, even if you haven't had direct experience with the object in question. An Asian may
have never eaten a slice of cheese but will still believe that it
smells bad. We have also implicitly learned the norms for basic safety and health through cultural transmission. You don't
have to have been trapped in a burning house to know that the
smell of smoke means fire. Learning that smoke signals danger
is sufficient.

"SMELL ARITHMETIC"

The British intellectual Sir Francis Galton (1822–1911), known for many eminent distinctions, including being the half cousin of Charles Darwin, decided to do a little experiment and taught himself to do "smell arithmetic." Galton associated specific smells with specific numbers;— for example 2 whiffs of peppermint = 1 whiff of camphor— and claimed that he could add and subtract quite well with imaginary scents, but that multiplication was too difficult.

EXCEPTIONS TO THE RULE

Despite all the evidence for odor-associative learning, a test ruling out the possibility of an innate response to *some* aroma or another, such as Grenouille's magical attractant elixir in the novel *Perfume*,[8] has not been done. Therefore, the theory of learned olfactory response is not irrefutable. Moreover, there are two physical factors that influence odor hedonic perception that also qualify odor-associative learning: trigeminal stimulation and genetic differences. Trigeminal stimulation refers to the fact that almost all odors have a *feel* to them as well as a smell. For example, menthol feels cool and ammonia feels burning. What produces these feelings are the temperature, touch, and pain fibers of the trigeminal system in our face and nose. Odors vary in the degree to which they stimulate the trigeminal system. Some, such as rosy floral scents, do so very mildly and are hardly felt at all, while others, like skunk spray in your face, can be powerful enough to make you cry. Indeed the reason your eyes tear when you chop onions is because of activation of the trigeminal nerve, and this is why skunk spray near your eyes would be immediately repellent. The trigeminal system is also why you sneeze when you sniff pepper and why your mouth burns when you eat habaneros, and it is even involved in the intense pain of migraine headaches. "Smelling salts" are actually a heavy dose of trigeminal stimulation and are made from ammonia combined with eucalyptus oil. The trigeminal irritation from ammonia and eucalyptus is what revives a person from a stupor.

Trigeminally irritating odors that elicit pain concomitant with odor perception induce immediate avoidance responses and appear to us as instant scent disliking. However, this dislike-withdrawal from the odor is due to chemical pain becoming associated to the chemical's scent. Skunk spray "scent" isn't painful, but its trigeminal component is. Instantaneous withdrawal from stimuli that are trigeminally irritating is adaptive for us, as many toxic substances strongly activate a trigeminal response. It would be fascinating to do an experiment in which the trigeminal component could be subtracted from a chemical like ammonia to see how liking for the pure *odor* of ammonia might change.

A second consideration is the individual differences in our genetic makeup. Variability may exist in the specific genes that code for olfactory receptors between individuals. It is quite likely that you and I don't have the exact same olfactory receptors in action. We also know that some people suffer from "specific anosmias"—the inability to smell one particular scent with otherwise normal olfactory function. Because individuals can differ with respect to the functioning of their olfactory receptors, biological (i.e., inborn) factors could be an influence in olfactory *perception*. Therefore, people who like the smell of skunk may do so in part because they are missing receptors for detecting some of the more pungent volatiles of its bouquet. Or because some of their receptors function differently, yielding a slightly different olfactory experience from those of people who insist that this scent is unquestionably offensive. In support of such a conjecture, Chuck Wysocki of the Monell Chemical Senses Center has found that identical twins always have the same immediate hedonic response to the smell of cilantro (also known as fresh coriander or Chinese parsley), either loving it or hating it, while fraternal twins do not. However, one of the snags with this finding is that the twins' past history with cilantro was not obtained. Identical twins tend to have more experiences and environmental influences in common than fraternal twins. Therefore, identical twins are likely to have shared a common past experience with cilantro, such as getting sick after eating a Mexican meal, and learning may still be at the root of this finding. Apart from this example, there is no evidence as yet for how or whether genetic differences might influence odor perception.

NATURAL HISTORY

Associative learning is the best explanatory model for our odor penchants. But why would it be that our responses to odors are learned and not innate? Does this make sense from an adaptive perspective? In other words, is it good for the survival of the species?

As mentioned earlier, "specialist" species live in restricted and specific habitats, eat only a few types of foods, and are prey to only a small set of local predators. The panda bear with his bamboo exclusive diet is one extreme example. From an evolutionary perspective, it would be adaptive for specialists to have hardwired responses to particular odors. If the panda didn't know that only bamboo was food, he would die from eating the wrong things. Evidence for specialist approach-avoid behavior has been found in a number of studies of animal behavior. For example, both lab-born and wild-reared ground squirrels show a discriminative defensive response to their natural predators, rattlesnakes, as compared to gopher snakes. This discrimination is made on the basis of subtle scent cues that differentiate the two snakes. The fact that this specific behavior is seen in both lab and wild squirrels suggests that their olfactory responses are innate.

In contrast to specialists, humans, along with rats and cockroaches, are the world's most successful "generalists"—animals that can exploit any habitat. If we were hardwired to only accept fishy smells as food, we would never have survived in the savannah. Thus for generalists, the adaptive response is to learn how to respond appropriately to a particular smell source when it is encountered, and not to follow a preprogrammed set of responses to particular odors. This is also important because the relationship between an aroma and a food or predator source can be random. From an evolutionary perspective it is adaptive for animals who are specialists to have innate olfactory responses to prey and predators, whereas animals who are generalists, such as us, should not be hardwired to know particular odors as food or poison or friend from foe. Rather, we should be hardwired to rapidly and readily learn what scents signify good and what scents signify bad, based on experience.

One of the best natural examples of adaptive and very rapid odor learning is the case of *taste aversions*. Being sick just once after ingesting a certain food causes avoidance of the substance that triggered the illness, and especially its scent, for a long time thereafter. You may be familiar with your own personal story of such an occurrence. Here's mine. One night when I was about nine years old my parents were going out for the evening and my brother and I were allowed the special treat of pizza for dinner. My parents decided to choose pepperoni. I had never had it before but knew it was the top choice of my classmates and eagerly awaited the delivery. Minutes after the doorbell rang, my brother and I devoured the pizza with delight. Later that night I began to have stomach cramps, then diarrhea and vomiting, accompanied by a fever, and these symptoms persisted for several days (my brother was fine). Although I well understood the diagnosis of stomach flu and that the pizza hadn't been poisonous, I could not bring myself to go anywhere near pepperoni pizza for a very long time. Indeed, for years after, the smell of pepperoni pizza would drive me out of a room. And to this day I find myself bizarrely reluctant to eat a slice of pepperoni pizza if it is offered. No other pizza topping ever brings up that feeling of caution; indeed BBQ chicken pizza is one of my favorite treats.

Intellectually knowing that it isn't a particular food that has made one sick does not override the instinct to avoid something that has been paired in time with getting sick; specifically, stomach illness. The "irrational" override of intellectual knowledge is based on our primeval past. Our ancestors were more likely to be eating a poisonous mushroom when they vomited after a mushroom feast than coincidentally getting the stomach flu. And you don't want to keep repeating the mistake of being poisoned. Rather, you should quickly *learn* that mushrooms with *that* smell should be avoided!

For generalists, as humans are, pepperoni or mushroom are not inherently meaningful smells in themselves; rather, their association to pleasure or pain is what then makes them interpreted as good or bad in the future. For generalists like us, it is evolutionarily adaptive that the olfactory system is *not* predisposed to like or dislike any particular odors, but rather very ready to become associated with what is good and what is bad based on experience.

The neuroanatomy of our sense of smell also shows how odors are inherently primed to become connected to emotional meaning. The piriform cortex, comprising the olfactory and limbic structures, is the area of the brain responsible for processing smell, and it is also the area of the brain where assigning positive or negative value to things takes place. Furthermore, the amygdala, which is directly connected to olfactory processing, is critical for emotional associative learning. In studies with another very successful generalist, the rat, it has been found that when the amygdala is eliminated, olfactory learning does not take place. A rat cannot learn the difference between a nutritious lemon drink and a poisonous strawberry drink without its amygdala.

If we have any innate response to odors, it is caution. Infants and young children show wariness when exposed to unfamiliar scents, regardless of whether these odors are classified as pleasant or unpleasant by the adults around them. This uneasiness in the face of uncertainty is adaptive. Without the predisposition for caution, our ancestors would not have sur- vived. The reason we each like or don't like the scents in our world is because of our specific personal and cultural histories with these scents and how we then thusly characterize and connote them. *Nothing stinks, but thinking makes it so.*

PSYCHOLOGICAL TWISTS IN ODOR PERCEPTION

Learning provides the mechanism by which we come to know an odor as good or bad, but it is not the only factor that influ- ences our perception of scents. Trigeminal stimulation and our genetic makeup may intercede, and as you might expect, psy- chological factors can play a role as well. Simply being in a good or bad mood can affect how pleasant you will find a smell to be. If you are in a good mood, a neutral* odor, such as rubbing al- cohol, smells more pleasant than if you are in a bad mood. This is yet another example of the fundamental connectivity and interdynamics between emotion and olfaction.

Our personality predispositions also lace our responses to smell, as they do our responses to all sensory stimuli. People with emotionally unstable personalities, sometimes called *neurotic,* tend to be more sensitive to noise, pain, unpleasant scenes, and bitter tastes than emotionally stable individuals. These emotionally

---

*"Neutral" refers to the averaged rating of an odor being in the middle of a scale anchored by "very pleasant" and "very unpleasant." It does not mean that all people find the scent to smell neutral.

excitable types are also more responsive and sensitive to odors, finding them more delightful or disgusting than their calmer counterparts. However, some studies suggest that it depends on what the odor is and whether you are a man or a woman.[9]

I have already alluded to the fact that language plays an intricate role in our perception of scents. Its force can be subtle but also powerful. As you now know, skunk odor is particularly interesting to me, and so I recently conducted an informal survey among my American colleagues to see how they reacted to this scent and how they *described* it. I was amazed to find that in addition to the great variability in liking, no two people used the same words to describe what skunk smelled like to them. I heard everything from "it smells like lemonade" [*sic*] to "it smells like sweaty socks." My own personal description, which no one else seconded, is that it smells like "a mixture of chocolate and garlic." Does what I call chocolate-garlic actually smell the same as what another person calls sweaty socks or lemonade? At this time, we have no way of knowing.

This linguistic issue illustrates one of the central problems in our psychological understanding of odors. We don't learn about odors the way we learn about other objects and experiences in our world. Our parents and teachers very rarely give us any "smell and tell" lessons. Therefore, most of us have acquired idiosyncratic connections between scents and words. Even if you did receive various declarative labels for assorted odors, your name-odor connection may still be relatively random. Suppose one day when you were four years old, you were walking by a steam grate with your grandfather and he said "oh, that smells like laundry," or he could have said "that smells like chlorine," and then again he might have said "it smells like steam." Whatever word was supplied will go on to be the label you use to refer to that scent in the future, unless you happen to be corrected, which is unlikely. So you can see how the word that is attached to a scent can be arbitrary and knowing whether it is the word or the perceptual experience that is different between individuals is currently an empirical conundrum.

Our expectations and the situational context we are in can also drastically influence what we construe an odor to be and therefore what hedonic properties we assign it. If you are walking by a garbage Dumpster and encounter a particular aroma with a few sharp notes, you are likely to perceive the smell as unpleasant; however, that exact same mixture of chemicals hovering above the cheese plate at a French restaurant might inspire salivation. Mark Twain gives a striking literary anecdote to illustrate this point in *The Invalid's Story*.[10] In this short story, a stowaway takes actions that eventually lead to his untimely death, because he convinces himself that the sack beside him contains a dead body, when in fact all it contained was "a lot of innocent cheese." In this situation, a suspicious context, and a burlap sack of a particular but ambiguous shape, created a set of expectations that influenced odor perception and even inspired drastic behavior.

A real-life example of how visual context can dominate in odor perception that is well appreciated by the flavored-beverage industry is the *lack* of complaints that are received when the occasional color-flavor error occurs. For example, if the purple-colored "grape" drink is accidentally flavored with cherry (not grape), the phone almost never rings. A wonderfully irreverent test of this same type of color deception even conned French wine experts. When the experts were given a white Bordeaux that had been adulterated with red food coloring to taste, they described it exactly as they would a fine red wine. And when the same white wine was given to them without any added coloring, they used all the descriptors they normally reserved for a lovely white Bordeaux. Helga Griffiths, a German conceptual artist who specializes in creating sculptural installations with scent, recently told me that casting a green light over a museum display emitting a grassy odor induced spontaneous remarks from spectators such as "this smells like grass," but when a red light was shone over the same display the scent provoked remarks like "this smells like strawberry."

L IKE VISUAL CUES, language also has powerful and unusual control over odor perception. My laboratory found that we could produce olfactory illusions by merely changing the verbal labels that were provided when someone smelled an odor. We tested five different odors and found that when we gave the same odor either a positive or negative name, such as, "Parmesan cheese" in one case and "vomit" in another,* we could induce entirely different perceptions of the scent and in essence create olfactory illusions. Not only did people react entirely differently to the odor based on its label, either saying they liked it very much and would eat it (in the Parmesan case) or that they were disgusted by it and wanted to leave the room

*For the mixture of iso-valeric acid and butyric acid.

(in the vomit case), they also would *not* believe that it was the same odor that they were smelling on both occasions.[11] So even though we may have initially learned that a certain scent is good because it is a tasty food, like cheese, that exact same scent can be turned into something bad simply by the associations evoked by the situation you are in, the words used to describe it, or the expectations you have while smelling it.

Why are we so easily deceived by words and visual context when it comes to the perception of smells? The reason is because scents are invisible and we are obsessed with identifying the objects in our world, and so with enigmatic smells we look to words and scenes for help. If the context offers a plausible explanation, we will go along with it to a greater extent than relying on the smell alone for what it is and whether or not we like it.

Not all odors are equally susceptible to illusory suggestion, however. The ease with which our noses can be deceived depends on how ambiguous the scent is. If I gave you *orange* to smell and told you it was "tangerine," you would likely believe me, but you might be a little skeptical if I called it "lemon," and you certainly would raise an eyebrow if I called it "garlic" or even a different type of fruit such as "banana." This is because the smell of orange is well rooted verbally and perceptually—it is not ambiguous. Odors with clear and obvious sources are less vulnerable to verbal illusory tricks. But odors that could be one thing or another depending upon the context you find them in can be perceptually twisted and turned in myriad directions.

Words not only create aromatic illusions; the absence of words can make real odors unsmellable. Several years ago, researchers in Britain did a test where an unusual odor, 5-alpha-androstan-3-one, was pumped into the air of a laboratory and those exposed to it were monitored for any signs of detecting it. The researchers found that participants produced physiological signs that were consistent with having detected a scent, but when asked if there was any smell present, most of them denied smelling anything. Later when the same individuals were given the odor and told its chemical name, those who denied perceiving it earlier suddenly recalled having noticed it in the room.[12] Words can manipulate what it *is* that we smell and, when no words are available, we may not smell a real scent.

In all languages that have been studied, there are fewer words that refer exclusively to the experience of fragrance than there are for any other sensation.[13] In English, *stench, stink, redolent, aromatic, pungent, fragrant, smelly, odiferous,* and *scented* exhaust the dictionary of words that specifically describe odor experiences. More common terms like *floral* or *fruity* are references to the odor-producing objects (flowers and fruits), not the odors themselves. We also borrow terms from other senses—chocolate smells *sweet,* and grass smells *green*—to describe our aromatic encounters. But we do not need words to experience olfactory sensations or to know how to react appropriately to them. Our experience of aromas can reside in a pure wordless smell-scape, and in fact is often purer and more exquisite when it does so. Wine connoisseurs begrudgingly admit that ignorance can be more blissful than the weight of their knowledge. Having a rich vocabulary to describe the flavor nuances of a vintage can actually diminish the richness of the experience, because it forces the expert to dissect and analyze a sip into parts that are much less pleasurable than the whole.

The situation or context we are in is also important because it determines what odors we will accept where. Certain smells, like certain colors and certain sounds, go together with certain things; yellow goes with pineapple, rings go with telephones. But this is not an innate or absolute truth. What sights, sounds, and smells go *best* with what objects is based on learning. I learned that shampoos had sweet scents in certain perfume and outdoorsy categories. And although there are many other sweet smells that I like, chocolate chip cookies for one, because of my early education I wouldn't want cookie dough aroma in my shampoo. Just as with the individual vagaries in the acquisition of odor hedonics, conviction in the congruence between smells and things can be quite idiosyncratic and is often a mystery to the smeller herself.

Marilyn Singer, the children's book writer, recently sent me an e-mail asking why she likes floral-scented candles but not fruit-scented ones, even though she likes both floral and fruity aromas. I wrote back that my guess was that when she first experienced scented candles they were floral not fruit scented, and as a consequence she *learned* what scented candles *should* smell like. She likes floral candles because they smell "right"—what scented candles ought to smell like (for her)—and she dislikes fruity candles because they smell wrong. In the case of fruit-scented candles, the scent and object are not congruent for her.

And how is it that our likes and dislikes of odors stay the same over time? You like the scent of lilac and dislike cigarette smoke time and time again. The same associations of happiness or disgust are elicited each time. Or are they? Did you once find cigarette smoke appealing, say when, or if, you were a smoker, but now as a nonsmoker find it offensive? Or maybe you disliked cigarette smoke as a child, but it became a scent marker for your father and it now elicits wistful and pleasant connotations. Intrinsic to the fact that we form associations to odors that determine our hedonic responses to them is that an association is a type of memory. An association can be vague and you may only *feel* that a certain smell is good or bad, with no specific recollection in mind, but it can also bring forth complex and intense personal memories. The association that made the scent of rose unpleasant for the woman who first experienced it at her mother's funeral is also a memory of her mother's funeral every time she smells it. And more than any of our other sensory experiences, smell is exceptional in its ability to conjure emotional memories and viscerally transport us in time and place.

CHAPTER 3

SCENTS OF TIME

Wherever I am in the world, all I need is the smell of eucalyptus to recover that lost world of Adrogue, which today no doubt exists only in my memory.

—JORGE LUIS BORGES

One day my cousin Amanda visited some friends whom she had not seen in a long time. The trip involved a considerable drive and it was decided that Amanda would stay the night. In return for her friend's hospitality, my cousin insisted on doing the dishes after dinner. As she bent over the soapy water and began the mundane task of scrubbing, Amanda was suddenly overwhelmed by inexplicable emotion. As if out of nowhere a bullet had hit her—so intense were her feelings that she found herself weeping. For several minutes her head hung over the sink as tears streamed down her face; she felt ridiculous, confused, and overcome with a strange nostalgic sadness all at once. Her friend watched, stunned and concerned; *"What's wrong? What's wrong?"* she kept asking. Suddenly, Amanda looked up, turned to her friend, and said: "*It's my grandmother. It's the smell of the dish soap. . . . I can see her perfectly. Standing in her kitchen and I'm helping her do the dishes the last Thanksgiving we all spent together. I'm right back there with her. I can't believe how much I feel her now. I miss her so much.*" My cousin continued to talk to her friend about her memories of our grandmother for the rest of the evening.

Has anything like this ever happened to you? If not with the scent of a certain soap, then with a special perfume, the aroma of school-floor varnish, the dusty odor from an old book, or some other distinctive scent? Although uncommon compared with other memory experiences, almost everyone I have ever met has experienced a startling and poignant memory that was triggered by a particular smell. And everyone who has ever talked to me about their scent-evoked memories has been fascinated and curious about this experience.

Throughout history, poets and philosophers ranging from Aristotle to Borges and Nabokov have pondered and extolled the wonders of memories triggered by scent. The most frequently cited example, probably because it is the most detailed, is the one given by Marcel Proust at the beginning of his seven-volume opus on memory. In the first chapter of *Swann's Way,* Proust recounts an episode where the aroma arising from a madeleine biscuit soaked with linden tea brings back a stunning long-forgotten recollection:

> No sooner had the warm liquid, and the crumbs with it, touched my palate than a shudder ran through my whole body, and I stopped, intent upon the extraordinary changes

that were taking place. An exquisite pleasure had invaded my senses, but individual, detached with no suggestion of its origin. . . . Whence could it have come to me, this all powerful joy? I was conscious that it was connected with the taste of the tea and cake but that it infinitely transcended those savors.[1]

From the fame of this description, these redolent memories are often dubbed "Proustian memories" or an example of the "Proust phenomenon." Proustian memories are typified as emotionally vivid, sudden, autobiographical recollections triggered by a scent. Odors have also earned a reputation as being the "best" cues—reminders—to uncovering our past. But are odors really the best cues to memory? And if so, how are they better than or different from any other cues at eliciting recollections?

THE BEST CUES TO MEMORY

In scrutinizing Amanda's and Proust's memory descriptions you may notice several similar features. First, pure emotion is the most immediate and central experience in these recollective episodes, and the content of the memory—what happened, who was there, where it was—becomes filled in only later. Another striking feature is that these memories are sudden and unexpected. We are usually startled that scents are capable of unleashing such powerful feelings and recollections because we generally ignore our sense of smell.

Most memories of our personal past are triggered in the same basic way. A feature of the original event is encountered and the event to which it was connected tumbles back. With Proustian memories, the feature that triggers the recollection is a scent. In Amanda's case we can trace the process as follows: Amanda had been very close to our grandmother who had died some years before. The dish soap that our grandmother used was uncommon, and Amanda had not smelled it since the last time she washed the dishes with Grandma. When Amanda unexpectedly came across the scent of this specific dish soap, the particular memory that was tied to it, our grandmother, was pulled out of the sink and into her consciousness. For Proust, we can presume a similar sequence of psychological events. Proust had not enjoyed the flavor of linden (lime flower) tea with madeleine biscuits since his childhood summer with his aunt, and then when many years later this mixture reached his palate, the memory to which it was linked erupted. These sequences show that a specific scent is the triggering cue to a significant past moment, but the question still remains: are odors actually *better* at triggering recall than things we might hear, see, feel, or taste?

Despite our inherent intrigue with scent-triggered memories, this phenomenon has been barely touched by science. The first "empirical" investigation of scent-evoked memories was reported in 1935 by Donald Laird, who summarized the work of a Mr. Harvey Fitzgerald at Colgate University and published it in the journal *Scientific Monthly*.[2] Fitzgerald interviewed 254 "men and women of eminence," including writers, doctors, and clergymen, and asked them to retrospectively describe what their scent-evoked memories were like. One woman reported a scent memory experience as follows:

On the train once, in the midst of happy conditions, I suddenly felt discouraged, awkward, unhappy. As soon as I recognized the perfume used by a fellow traveler, I saw very vividly a large dancing class, a French dancing master, and felt again my girlish dismay at his attitude toward my poor attempts to learn the steps he was trying to teach me. As soon as the memory picture came I knew why I had suddenly felt unhappy, and, of course, came back to normal. This experience occurred some fifteen or twenty years after the last time I had seen the dancing master.

Further attesting to the claim that odors are especially effective memory cues, one of the interviewees, a midwestern bishop, remarked that "smell is sure some automatic reminder." An overall summary of these descriptions led Laird to conclude that scent-evoked memories are markedly vivid and emotional, and that smells often trigger memories of long-ago events. The problem, however, is that this study does not go beyond being a compilation of anecdotes itself, nor does it address the comparison between scent and other memory cues.

Fifty years later an attempt was made to bring scent-evoked memories into the laboratory. In 1984, David Rubin and his colleagues Elisabeth Groth and Debra Goldsmith at Duke University conducted a study on autobiographical memory where they compared memories elicited by items presented either verbally, as photographs, or as odors.[3] If a memory was triggered by an item, in any of the sensory forms, the participant was asked to rate his or her memory for the qualities of vividness and emotionality, and for how often the person had thought of or talked about the memory prior to the experiment. Rubin and his colleagues found that memories elicited by items presented as odors were reported as *thought of* and *talked about* less often than memories elicited by items presented as photographs or words. In other words, scent-evoked memories tended to be of rarely relived events. But their findings did not substantiate the conviction that scent-evoked memories are more vivid or possess any other exceptional qualities, or that they are in any way "better" than memories triggered by visual or verbal cues.

THE EMOTIONAL DISTINCTIVENESS OF SCENT-EVOKED MEMORY

In 1990, I began to chase the question of whether and how odors might be the best memory cues, and ever since then I have been investigating the nature of scent-evoked memories with a variety of techniques.

My first challenge in this endeavor was to decipher what people meant when they talked about a "good" memory, or "best" memory cue. Typically we think of a "good" memory as being an accurate memory, a vivid and correct recollection of an event as it actually took place. But memory is more than just an accurate mental presentation of the past. In addition to the facts, like remembering where Grandma lived, memories have a personal, subjective, and emotional dimension.

Recollections of our past are always accompanied by a feeling, which can range from a vague nostalgia to poignant and intense emotion. Because of this dimensionality, I parsed memory into two components in order to study it: its *objective accuracy,* who was there, what they were wearing, what someone said; and its *emotional quality,* the feelings and evocations that arise with the return of a past experience.

To test the differences between scent-evoked memories and other types of memory, I developed a procedure similar to that used by Rubin and his colleagues and compared memories triggered by a cue presented as a smell with the same item presented in other sensory forms. For example, the recollection stimulated by the *smell* of popcorn would be compared against the memories evoked by the *sight* of popcorn, the *sound* of popcorn popping, the *feel* of popcorn kernels, or simply the word *popcorn.*

What I found in these experiments is that, in terms of their accuracy, detail, and vividness, our recollections triggered by scents are just as good as our memories elicited by seeing, hearing, or touching an item—but no more so.[4] Yet our memories triggered by odors are distinctive in one important way: their emotionality. We list more emotions, rate our emotions as having greater intensity, report our memories as being more emotionally laden, and state that we feel more strongly a sense of being back in the original time and place when a scent elicits the past than when that same event is triggered in any other way. I have also found that the amygdala, the wellspring of emotion in our brain, is more highly activated when a person recalls a memory by the scent of a perfume than when a person recalls the same memory as a result of seeing that perfume's bottle, or when he or she sees or smells a nonmeaningful perfume. So scent-evoked memories are *different* from other types of memory experiences. They are uniquely emotional and evocative—in our minds and in our brains.

CAPRICIOUS EMOTION

A memory conjured by an aroma will always feel emotionally intense, but the emotions experienced with that memory are not fixed. A certain scent may always remind you of the same person or moment and always be accompanied by a rush of feelings, but though the content of that memory will stay the same over a lifetime, the sentiments carried with that reminiscence may not.

Suppose that when you were sixteen you were completely infatuated by a popular girl in your class—let's call her "Nancy"—and to your astonished delight she actually said yes when you invited her to your high school dance. Gleefully, you soon found yourself Nancy's boyfriend and whenever you were with her you noticed that she wore a lovely, distinctive perfume, one you would later come to know as Chanel No 5. During those first blissful months, whenever you stumbled upon Nancy's scent you became suffused with passionate diversion. Then, after several months of heavenly dating, one night at a large party with all your jealous friends in attendance, Nancy unceremoniously dumps you. Your agony, humiliation, and anger are insurmountable. Now when the scent of Chanel No. 5 comes into your nasal view, terrible emotions come racing with it, a million miles from the blissful feelings that just days ago accompanied her fragrance. Fast-forward fifteen years into the future when you are happily living with your long-term partner, confident and contented; one day on your way to work a woman passes you on the street wearing Chanel No. 5, and you find yourself in a laughing reverie. You remember Nancy all right, but instead of bliss or humiliation, a very different emotion is brought with that reminiscence. Now the forgiving brush of nostalgia has painted your feelings with humor and wisdom.

This vignette shows how the emotions that come with your memory of Nancy, though always strong, will not always be the same; they depend on what Nancy means to you at the time you are remembering her. The emotions that accompany specific memories change as a function of what the event means to you at the time you are remembering it. This emotional instability is not restricted to scent-evoked recollections, but the intensely emotional nature of scent memories makes them an especially revealing display of this effect.

Despite experiences like the Nancy story, scents are no better than other kinds of memory triggers in terms of the accuracy of information they elicit, but they are uniquely emotionally involving. I believe the distinctive and intense emotionality of scent-evoked memories offers a critical insight into why odors are regarded as the "best" memory cues. It is *because* of their intense emotional evocativeness. The emotional potency of odor-evoked memory leads to the false impression that these memories are especially true, and that odors are superior reminders of our past experiences.

Scent-evoked memories provide an excellent illustration of the persuasion of emotion and also a cautionary reminder. The confidence that one's recollections are accurate, which is so hard to resist when memories are colored by emotion, is similar to what often emerges on the witness stand. Eyewitnesses tend to be doggedly confident that their recollections are correct. Unfortunately, research shows that their memories are often quite mistaken. However, the misconstrual of emotional intensity for accuracy may not be the complete explanation for why scents have earned the reputation as superior reminders. There is another feature of scent-evoked memories that Proust alluded to which may truly make odors better than other memory triggers. The Canadian writer Anne Mullens recently told me about an event in her life that vividly captures this special quality.

MEMORIES LOST AND MEMORIES REGAINED

At the age of nineteen Anne was touring a veterinary college with some friends when they wandered into an empty room with a large drainage hole in the middle of the floor. Despite the barrenness of the room, Anne found herself riveted in place as soon as she entered. A scent emanating from the hole, an odd mixture of what Anne described as animal blood, sawdust, and cleaning fluid, released the floodgates on a long-forgotten event—the memory of being six years old and smelling a very similar concoction during a tour of a slaughterhouse in a remote Newfoundland village, where her father was the visiting doctor for the summer. According to Anne, "smelling the odor was like pressing play on a lost movie in my head." In telling me this story, Anne laughingly recalled that her friends thought she was having a seizure because "I could not move from the room and was staring off into space with my mouth wide open as if being fed this memory from out of the blue." Anne's voice quickened as she launched into the rest of her memory:

> I recalled being six in a pig slaughterhouse in the hamlet of St. Anthony, at the northern tip of Newfoundland. Since my dad, the visiting doctor, was the local dignitary, they showed us all the sites, including this slaughterhouse. This chunk of memory that came back included the whitewash-painted wall, the gray cement floor and the white wood pillars, and a table with a big sink against a wall. The wall had some blood splatter on it.
>
> After the tour we went to a family's home—the full-time doctor's house—and sat in their living room. I can see the lighting of it and the very formal look—green velvet chesterfield, chintz chairs, walnut coffee table, and side tables. Very English and formal for a small fishing village. We were served tea and I remember lemon pound cake and rosebud china and that the woman, Mrs. Thomas, was dressed in a nice floral dress. I think I was wearing a floral dress, but I have lots of memories of a favorite dress, the strawberry dress (white with red strawberries), and I can't be certain if I was indeed wearing it, as it is in a lot of memories and not unique.

Awestruck by having such an exceptionally vivid lost piece of her past thrust back into her life, Anne immediately telephoned her mother to verify that her memory was real. To

Anne's delight, her mother confirmed that they did tour the local slaughterhouse during that summer in northern New-foundland, and that after the tour they had all gone to the home of the Thomases, the local full-time doctor. But she could not confirm any other details that Anne remembered of that day, because she could not remember them herself. Nor could Anne's three sisters. Anne laments that she cannot judge the accuracy of her memory, since no one else can remember it well enough to corroborate her recollection. But that does not diminish the fact that this memory felt extremely real and was ". . . a truly phenomenal experience which has fascinated me for years."

This anecdote captures a special dimension of the Proust phenomenon, the ability of odors to reawaken lost memories. Is it possible that odors have the ability to trigger memories that might otherwise be forever forgotten, that we would never re-cover if not for stumbling upon the "right" aroma? This ques-tion has not yet been directly tested, but that this could be so is supported by two well-studied psychological phenomena: dis-tinctiveness and interference.

A distinctive object stands out from the background either because of its rarity in time, type, or both; when something is distinctive, we pay more attention to it. A particular smell may be encountered just once in your lifetime, or only in tandem with a very specific event, and so becomes forever linked to that single event; visual or verbal versions of the same cue, however, are much more likely to be repeatedly encountered and as such lose their distinctiveness and meaning. Proust noted this in the following passage:

The sight of the little madeleine had recalled nothing to my mind before I tasted it; perhaps because I had so often seen such things in the interval, without tasting them, on the trays in pastry-cooks' windows, that their image had dissociated itself from those Combray days to take its place among others more recent.[5]

The same phenomenon occurred with my cousin Amanda. The distinctive smell of a particular brand of dish soap was the trigger to remembering our grandmother, because no other dish soap smelled like it. In contrast, *seeing* various dish soap bottles or *hearing* brand names—including the type used by our grandmother—were impotent at evoking our grandmother because Amanda had encountered facsimiles of these sensations so many times before.

Another factor that allows odors to remain faithful memory cues is the comparatively low likelihood of encountering them. In Anne's case, that strange brew of animal blood, sawdust, and cleaning fluid was only experienced during her childhood trip to the slaughterhouse. Similarly, Proust must not have dipped a madeleine biscuit into linden tea in the years intervening between his summer as a youth in Combray and when he wrote his famous recollection in his late thirties.* Indeed

---

*I am assuming that Proust was in his late thirties when he had his famous scent-evoked memory, because *Swann's Way,* where the quote appears, was first published in 1912 when Proust was forty-one, and it is reported that he began work on *The Remembrance of Things Past* (*Swann's Way* is the first volume) in 1909.

the more uncommon the scent, the more likely that it will be associated with a unique episode from our past. How often is a deeply significant memory revived by the smell of coffee?

Not only are odors less frequently encountered than visual or auditory cues, but when an association is made to a scent, it is actually much harder to undo and reassociate it to another experience than it is for visual and auditory items to be reconnected and associated. In other words, the first association made to an odor *interferes* with the formation of any subsequent associations. A familiar example is the experience of "learned taste aversions." Coincidentally becoming ill after eating pepperoni pizza led to my severe aversion to the smell of pepperoni that persisted for a very long time and was extremely hard to unlearn, even when I knew that pepperoni was never to blame for my nausea.

Resistance to being overwritten and the high distinctiveness of certain scents can combine to make odors faithfully and directly tied to particular events like no other cue, and thus the unique key to unlocking a special memory. Even though this may be the feature of scent-evoked memory that makes odors "better" memory cues, their emotionality is still their most extraordinary feature. I have often wondered, if a scent unleashed a memory that otherwise would have been relegated to the memory dustbin, would that memory be more emotional than if some very distinctive sight or sound reawakened the same event? My guess is yes. When an aroma triggers recall, you are caught in a wave of emotion and evocation like no other. It is not the "long-forgottenness," but rather the unique connection between olfaction, emotion, and memory that makes scent-evoked memories so special.

TRAUMATIC MEMORIES

The spellbinding rush of emotion and transport to another time and place when a scent returns you to a lost love, a childhood event, or even washing dishes can make the past appear more powerful than the present. But not all memories released by aromas are of lost summer days or old lovers; odors can also trigger memories that are severely traumatic in nature.

There is a psychological disorder defined by traumatic memories, called *posttraumatic stress disorder (PTSD)*.* When individuals with PTSD experience the memory of their trauma, they can become as emotionally overwrought as they were during the original harrowing episode. To the victim, a PTSD flashback of rape can feel as bad as the rape itself. In PTSD, memories are triggered by a cue that reminds the person of the trauma. For example, walking by a parking lot could trigger the memory of being raped in a parking lot, but scents are the most insidious and vicious reminders. The intensely emotional nature of odor-evoked memories and their unique neurological links with emotional processing mean that when a scent triggers a PTSD flashback, these episodes recapitulate a complete reliving of the devastating event.

------

*A person with PTSD is plagued by the experience of recurrent memories of a traumatic event. Rape, fire, car crashes, and war are common traumatic events that lead to developing PTSD. Not everyone who undergoes a traumatic event will develop PTSD; personality and other predisposing factors interact to determine whether PTSD will be established.

Another reason why scents are so menacing for victims of PTSD is because they have no way to prepare or protect themselves from a sudden attack. Unlike visual cues, which can be intentionally avoided—*don't walk by any parking lots*—odors are invisible and can manifest almost anywhere. The scent of the specific cologne worn by the rapist could "appear" while the recovering victim is innocently walking down a street, sitting in a restaurant or a movie theater, or in myriad other unexpected places, and more than any other reminder can bring on a sudden and unexpected attack with paralyzing power. Aromas connected to PTSD can also make once normal activities impossible. Sometimes called the *barbecue effect,* the aroma of meat on the grill can trigger horrific memories of burned bodies and make summer cookouts impossible for firefighters, rescue workers, or war veterans for years afterward.

Not only are scents the worst triggers for PTSD flashbacks, but they are also the hardest to treat. I have had calls from psychiatrists and clinicians who are at their wit's end about how to deal with the scent trigger to trauma with their PTSD patients. While it is possible to modify one's life by avoiding parking lots, how can one avoid encountering an offending smell without becoming a complete recluse?

In such cases, I suggest to therapists that they employ a variant of the standard technique for treating fear and phobias called *systematic desensitization.* Systematic desensitization works by exposing the traumatized person to very mild facsimiles of the scary stimulus, then gradually moving to greater and more realistic exposures, and often to the real thing in the end. At each stage of exposure the patient is taught relaxation techniques to use while being exposed to the frightening object or cue. In the case of a woman suffering from PTSD who was raped in a parking lot, an example treatment for the visual stimulus of a parking lot could be dealt with in a series of steps from very mild exposures, such as an architect's drawing of a parking lot to finally being taken to the parking lot where the rape occurred. Sounds associated to the trauma, like slamming car doors, could be dealt with in a similar way. But how does a therapist deal with the fear triggered by the rapist's smell?

Theoretically, smells should be able to be dealt with using systematic desensitization just like visual and auditory cues, but there are several difficulties. First, how do you come up with a mild facsimile for an odor? If the victim recognized the smell of her rapist as Old Spice, treatment could begin by talking about Old Spice, then seeing Old Spice bottles, working up to having the patient smell Old Spice. But visual and verbal versions of an odor memory cue do not have anywhere near the emotional potency of the odor cue itself. So although the concept of Old Spice might become neutralized, this will likely not translate into a neutral response to the scent of Old Spice. Second, the fact that scents inherently become more tightly glued to emotional associations than any other type of sensory cue, desensitizing—that is, unlearning—emotional associations to the real fragrance will be very difficult. Nevertheless, progressing through various sensory forms of a cologne and working up to repeated presentations of the real scent while using relaxation techniques and positive associations should at least be somewhat effective.

There is some evidence that using scent to treat PTSD patients is indeed helpful. Skip Rizzo, a psychologist at the

University of Southern California who treats Iraq war veterans suffering from PTSD, has recently begun incorporating smells into his therapy and has found that the addition of these scents makes treatment more successful. Dr. Rizzo uses a virtual reality video-game technique as a form of systematic desensitization. To make the video game more realistic, a range of odors, such as diesel, gunpowder, and garbage are emitted during various scenes to make the war veterans feel more like they are really back in Iraq. This more realistic immersion into the original traumatic episode makes overcoming the destructive nature of their memories more possible. As you will see later, scents are being used in many imaginative ways to make visual experiences more real.

A COLLECTIVE PTSD: THE SCENT OF 9/11

October 30, 2001, was a beautiful fall day in Brooklyn, and Nathaniel was distracted by thoughts of work as he entered the subway on his way uptown. As the train neared the Chambers Street station, in lower Manhattan, his mind suddenly flashed and froze on the present. It was *that* odor. A rush of panic pulsed through him, and looking around it was clear that the other passengers were similarly transfixed. *That* very distinctive odor, which lingered in "the zone" for months after 9/11, was known to almost all who experienced the collapse of the World Trade Centers firsthand, and whenever it was reencountered it served as an instant reminder of that day. Talk of the strange powers of the scent elicited by the destroyed

World Trade Centers led scientists Pam Dalton and George Preti, from the Monell Chemical Senses Center in Philadelphia, to analyze the air that permeated Ground Zero to see if anything unusual could be discovered. It turned out that the scent of Ground Zero comprised a complex and unique mixture of chemicals that smelled rubbery, bitter, and sweet at the same time. But, unlike the New Yorkers who lived through 9/11 firsthand and whose immediate response to this aroma was terror and grief, the Philadelphian researchers had no emotional responses to it at all. To Dalton and Preti, the aroma was merely strange—not bad and not good—yet the only difference between the Monell scientists and the New Yorkers was their past experiences with *that* scent. For the New Yorkers, the first time they had encountered this odor was a terrifying and traumatic world event, but to the Monell scientists this scent was merely associated with an interesting experiment. The response of the New Yorkers versus the Philadelphians epitomizes how the emotions associated to an odor will determine how that odor is later perceived and experienced.

SCENTSATIONAL MEMORY

If odors are such potent cues to memory, can they also help us remember? Could fragrances help you perform better on a test, or be used systematically to help you be better at recalling any sort of information you wished to? According to Simon Chu, an olfaction researcher at the University of Liverpool, Chinese culture has long been wise to the assistance of aroma in the retelling of stories. A Chinese tradition practiced for centuries was to pass around a small pot of spice or incense when generations gathered to share oral histories. Later when family members wanted to remember a story in detail, the same scent was passed around again.

The idea that odors might be used to enhance memory is supported by a well-established psychological phenomenon called *context-dependent memory*. When you are in the same context, place, or mind-set that you were in when you learned something, you remember that information better. It turns out that the key aspect of the context that helps you remember is how you *feel* when you are there.* Since odor is the sense most related to emotion, it would follow that odors would be the most effective cue to enhance memory. The idea that odors could be used as context cues has certainly occurred to people before me, and several researchers have already shown that scenting a room with a particular aroma and having that same aroma present during a test can sometimes improve memory— but not always. I was curious why odors were effective memory boosters only some of the time and so I conducted a series of experiments from which I discovered several things.

First, I found that in order for an ambient scent to facilitate memory, it has to be distinctive or unfamiliar—to stand out

*The discovery that one's feelings are the most important cue for context-dependent memory is credited to Eric Eich, a professor of psychology at the University of British Columbia.

from the background so that your attention will be drawn to it, even if you don't intentionally focus on it.[6] You would likely ignore the scent of cleaning solution if you smelled it in a laboratory, but your nose and mind would prick up if a sterile laboratory were fragrant with peppermint or butterscotch. The second thing I discovered is that emotion is a fundamental factor in this effect.

Learning and memory are two sides of the same coin. In order to remember something you must already have learned it, and an emotional context makes events more memorable. If I told you a boring story, you would be less likely to recall the details of it than if I told you a story with the same essential content but where an emotional charge was added to its meaning. Which story would you remember better: (1) A young blond-haired boy wearing a blue shirt crossed the parking lot and got into his mother's SUV; or (2) A young blond-haired boy wearing a blue shirt was struck by a speeding motorcycle as he crossed the parking lot toward his mother's SUV? I am sure you will agree that the latter is more memorable.

Considering the power of emotion to intensify memory in general and the particularly privileged connection between our sense of smell, emotion, and memory, I wondered whether a heightened emotional state experienced while an odor was present would hyperactivate the neural underpinnings of emotion, scent, and memory, thereby causing an odor and an event to become superglued together.

To test this theory, I conducted two experiments with forty-eight and forty students, respectively, in which an unfamiliar fragrance was present in a room while people *learned* a random list of sixteen words such as *ring* or *horse*.[7] The students were not aware that I would later test them for their memory of these words. Nevertheless, the procedure was geared to help them learn (and hence remember) the words by having them come up with an event that had happened to them for each word in the list. One week later, at the test session of the experiment, the participants were then quizzed for their memory of the sixteen words. The number of words that were correctly remembered was the measure of memory; the more effective any available memory cue was, the more words would be recalled.

The key to these studies was that one group was in an anxious mood during the learning session of the experiment and also had an ambient aroma available as a potential memory aid. To induce anxiety I took advantage of the fact that I was at a university and exploited a naturally occurring anxiety-provoking event—exams. The participants in the "anxiety" group were anxious because the experiment took place one hour before a midterm exam. Students who were in the "normal" mood condition—experiencing no strong emotions of any kind—took part in the experiment in the hour before a regular class. When memory for the word list was tested one week later, it was before a regular class, and everyone self-rated as being in a normal mood.*

---

*It should be noted that students self-selected to be in this experiment and the most exam-anxious students undoubtedly did not volunteer. Nevertheless, the mood of the students was measured in each condition of the experiment, and students who participated in the hour before their midterm were significantly more anxious than the students in any other group.

The results showed that students in the group who learned the list of words in the hour before their midterm exam and who had the same ambient odor present during both the learning and test sessions remembered more words than students in any other group. They remembered more words than students who had experienced the odor in room air both times but were always in a normal mood, and strikingly, better than the group of students who were anxious during the learning session but did not have an odor cue available during the test session.* In fact, students who were merely anxious during the learning session did quite poorly, presumably because they were distracted by their upcoming exam and not paying much attention to the word list. The reason this is noteworthy is because emotion usually intensifies learning, but it wasn't helpful for the random word list when an odor cue wasn't there. However, if an odor cue was available and a heightened state of emotional arousal experienced, in spite of distraction, the students were able to remember even irrelevant information well.

There is a useful take-home message from this research, especially for students, but also for anyone who is trying to learn or memorize information. If you are anxious or emotionally worked up while studying material that you will later have to remember, it would be a good idea to have an unusual or unfamiliar fragrance with you that you can also take to your

---

*A context-dependent memory effect was also seen in this study. The participants who experienced an ambient odor both times remembered more words than participants in the two groups who did not have an odor cue available at both learning and test.

test. This is a perfect way to cheat without cheating. Before you rush to find a suitable scent for your next memory-challenging situation, however, there are two things you need to know.

Your first concern is with a physiological feature of our sense of smell called *odor adaptation,* the fact that after about fifteen minutes of smelling a particular aroma you effectively no longer perceive the scent. An example you may be familiar with is the unfortunate disappearance of aroma that occurs in scrumptious-smelling environments, like a bakery or flower shop. When you first walk into a busy bakery, the delicious aromas of sweet baked goods engulf you, but after waiting in line, by the time you are ready to buy your cake, the heady bakery aroma seems gone. Another frustrating situation you may be familiar with is if you have ever tried to willfully conjure a lost love or poignant childhood moment by fiercely sniffing an old cologne bottle or sticking your head inside a cedar trunk. The more you keep inhaling to relive those lost moments, the more both the scent and the memory you are desperately trying to recapture fade further away.

In these scenarios a physiological phenomenon has occurred inside your nose. When your odor receptors have been bombarded with particular molecules for a certain length of time, they cease responding to those specific chemicals. This physiological effect can be undone relatively quickly by simply removing yourself from the proximity of those odoriferous molecules. After several minutes of standing outside the bakery or leaving the old cologne bottle, you will be able to return to those aromas and smell them again with full pleasure.

One way to prolong the effect of smelling a scent before adaptation kicks in is to dispense an odor intermittently. Rather than having the air freshener on constantly, bursts of air freshener alternated with no scent will draw out the time before your receptors get saturated and prolong your appreciation of the minty pine aroma in your car. In the case of studying, sniffing an odor sporadically throughout your study session, rather than constantly, will increase the time it remains useful.

The second idea in using aromas as memory aids is more practical. Smell a different fragrance for the different topics you are trying to learn. If you are studying for a calculus exam and a driver's license test, be careful to use different scents for each of these topics and to not get them mixed up. If you use the same scent for both, you might find yourself thinking of speed limits when you should be remembering rates for asymptotes.

REMEMBERING ODORS

Memory can restore to life everything except smells.

—VLADIMIR NABOKOV (*MARY*)

Ironically, though odors are exceptional triggers of memory, it is extremely difficult, if not impossible, to summon up the memory of an odor itself. You can remember what a crackling fire sounds like, or what the house you grew up in looked like, but can you truly conjure up the smell of your old camping tent? Or even chocolate chip cookies?

Research into our ability to remember odors was pioneered by Trygg Engen, who was briefly mentioned in the previous chapter.

Born in Oslo, Norway, Trygg came to the United States in 1948 and spent his professional career at Brown University. Among Trygg's major findings is that our ability to remember odors is very long lasting and that the first association made to a scent is very hard to undo. Yet, in spite of Trygg's groundbreaking work, research into our ability to remember odors is pierced with an impeding conundrum: how do you study the ability to remember odors? This is a problem because of another central question in the psychology of smell: can we recall, that is, imagine, fragrances? We may be able to recognize fragrances—you can recognize the scent of the white-and-red candy from the restaurant as peppermint—but can you truly capture the *scent image* of a peppermint in your mind's nose when the candy isn't there?

I once conducted a survey with 140 college students in which I asked them to try to conjure various physical sensations, such as visualizing a car, hearing an alarm clock, feeling satin, tasting a lemon, or smelling chocolate, and found that the reported ability to conjure the aroma of chocolate was weak and considerably worse than the ability to conjure any other kind of sensory image. Other researchers have also failed to find evidence for olfactory imagery, but there are still those who argue that odor imagery is equivalent to imagery in our other senses. Professional perfumers, wine tasters, and chefs tend to be especially convinced of their odor imagery prowess. But whether they can truly image scents, or whether they are any better than the average person, has never been tested.

The most compelling data supporting imagery with our other senses come from neuroimaging studies, which show that the same areas of the brain are active during imagining and perceiving a particular sensation; for example, seeing pumpkin pie and visualizing pumpkin pie light up the same regions of the visual cortex. This neurobiological overlap between imagining and true sensation has been well documented for vision and hearing, but the same is not true for perceiving and imaging odors. The parts of the brain involved in actually smelling pumpkin pie do not overlap neatly with the parts of the brain that are active when you imagine the aroma of pumpkin pie.

Another place to look for imagery is in dreams. Research on dreaming has shown that dreams containing scent experiences are extremely rare, much rarer than dreams containing any other form of sensation. We also cannot smell while we are asleep. That is, you really do *wake up and smell the coffee,* and not the other way around. In a study I recently carried out with Mary Carskadon, a world-renowned sleep expert at Brown University, we found that during deep and dreaming sleep,[*] even very strong and trigeminally activating odors such as peppermint and the harsh smell pyridine could not awaken sleepers or produce brain wave changes that were indicative of arousal.[†] None of our other senses are cut off so completely while we sleep. We do not currently understand why or how olfaction shuts down while we sleep. But one possibility is that because

---

[*]Deep sleep is slow wave stage 3 and 4 sleep. Dreaming sleep is rapid eye movement (REM) sleep. For more details, see Carskadon, M., & Herz, R.S. (2004). Minimal olfactory perception during sleep: Why odor alarms will not work for humans. *Sleep, 27,* 402–405.

[†]EEG (electroencephelogram) recordings were measured as an indication of neural arousal.

our sense of smell relies so heavily on interpreting context and paying attention and because this level of awareness is shut off while we sleep, this may be why our noses become ineffectual detectors when we are in this altered state.

For most people it seems that the sensation of re-creating the image of an aroma is derived from related perceptions and memories. Remembering what a turkey looks like, browned and gleaming, as it emerges from the oven; the warmth of the kitchen; a mood of happy satisfaction; and the anticipation of the savory flavors all conjure the *feeling* of smelling turkey at Thanksgiving. But our mind's nose is most likely not experiencing turkey the way we experience it when it is really being carved in front of us.

I believe the reason why we are so poor at imaging odors compared to other sensations is because we do not need abstract odor images to survive. We use smell to tell us what to approach and what to avoid when we come across an item in question— *this food is good and this one is bad.* We don't use odors to construct maps or abstract schemas of our world. Animals, like rodents, who rely predominantly on their sense of smell to negotiate the world, likely do think in smell, and certainly some of our primate ancestors did as well. But for modern humans, vision and hearing-language are the sensory information sources we use to construct abstract representations to make sense of and survive in the world. Because we don't rely on odor images in this way, this ability has not been specially selected for and hence has become weak. However, because selection processes are variable, certain individuals may still possess our ancestral imagery ability, and true odor imagers may be among us. Furthermore, because learning is such a central component of our sense of smell, it may be possible that with training or repeated experience one might be able to develop the ability to create sensory odor images. Some perfumers and chefs could be telling the truth and may indeed have acquired true sensory representations for the aromatics used in their craft.

Our memories make us who we are. Without memory we are at sea in a constant jumble of the present, lost from a world context and, most important, from ourselves. We do not know where we have been nor can we remember where we are going. Losing one's sense of smell does not destroy our memory, but it diminishes and alters it. The poignant and wistful feelings of nostalgia, the ability to conjure lost loves and certain long-forgotten events, are gone without a sense of smell. Jessica Ross complained that she no longer knew why she used to enjoy taking walks after the rain or visiting certain places. She said that without her sense of smell she felt disconnected from herself and other people. Ultimately she divulged that she thought a part of her was now missing, and that her sense of self and overall well-being were irrevocably damaged.

CHAPTER 4

AROMA AND THERAPY

The way to health is to have an aromatic bath and scented massage every day.

—HIPPOCRATES

When a person loses his or her sense of smell, like Jessica Ross, it can lead to unhealthy emotional states and a decreased sense of well-being. Could the converse be true, and the smell of aromas be curative? Can specific aromas be used therapeutically to improve our health, energy, wellness, and happiness?

A few years ago I asked Holly, one of my students, to do a little aromatherapy spying for me and find out what was happening behind spa doors. We arranged an appointment at a local beauty salon with aromatherapy services, and the following is Holly's account of her hour in this world.

When I arrived at the salon I was greeted by an attractive young woman looking official and medical in a white lab coat. She asked me to follow her to a very warm, dimly lit room at the back of the salon where soft, tinkling Asian-concept music was playing. There she instructed me to undress and lie down under the pink blankets of a narrow massage table in the middle of the room and said she would return shortly for my aromatherapy massage. When she returned, she adjusted the music, and I began to hear soft waterfall and rain sounds. She then asked me what my specific problems were. Professor Herz and I had discussed some general stress complaints to bring up, so I told her I felt anxious about school, was having trouble concentrating, and wasn't sleeping well. She nodded and moved to a table lined with small amber vials. I watched as she chose a selection from the large array and laid them on a tray beside the table. As she massaged the selected scented oils onto my back, neck, and shoulders she explained what each one was good for and how I could use them at home. Lavender was excellent for relaxation, and I could also put it in my bath. A few drops of sweet marjoram on my pillow would help me sleep, juniper berry was calming and helpful for overcoming mental fatigue, and sandalwood would help me be less stressed in general. She also suggested that I use these oils in teas or compresses, as well as in my bath and on my clothes and skin. After about forty-five minutes of this oily education, she announced that the treatment was over, and that I should get dressed and meet her at the front desk. I felt so relaxed that I just wanted to stay where I was and sleep. When I finally got to the front counter, my aromatherapy technician was waiting for me with bottles of the various elixirs she had used during my massage—clearly with an expectation of a sale. I told her that I would come back for them later, paid for my massage, and left.

Holly and I discussed her experience in detail, and combined with my research on the connection between scents, emotions, moods, and behavior, I am confident that I now know why people think aromatherapy has curative powers, and how it can in fact sometimes work.

WHAT IS AROMATHERAPY?

The concept of aromatherapy arises from ancient cultures that believed in the healing properties of plants. Aromatherapy was used by the Chinese as incense, by the Egyptians in embalming the dead, and by the Romans in their baths. Indeed, the intimate relationship between plant aromatics and medicine was likely forged long before recorded history as a way to ward off insects and treat skin conditions.

Many plants do have therapeutic properties, and the basic derivatives of many modern medicines are plant based. Acetylsalicylic acid, the active ingredient in aspirin, came from the discovery that chewing willow bark alleviated pain, inflammation, and fever. Oil distilled from thyme is estimated to be twelve times more powerful as an antiseptic than phenol, a common synthetic ingredient in cleaning agents. However, beyond basic antimicrobial and anti-inflammatory effects, contemporary aromatherapy proposes that various plant-based aromas have the ability to influence mood, behavior, and "wellness."

The term *aromatherapy* (originally *aromatherapie*) and its modern practice was launched in the mid-1930s by the French chemist René-Maurice Gattefossé, who began the exploration of essential oils for their healing powers after an explosion in his lab left his hand badly burned and the accidental soaking of his injury in pure lavender oil produced a rapid and miraculous healing.

Aromatherapy uses a wide range of pure and natural essential oils distilled from plants, shrubs, trees, flowers, roots, and seeds that are believed to possess healing properties. An essential oil is the distilled essence of a particular plant combined with a vegetable oil base. Different parts of the plant are used depending upon the flora. Geranium oil is from the leaves and stalk, bergamot oil from the peel of the orangelike fruit, and cinnamon oil from tree bark.

Aromatherapy practitioners believe that the scents distilled from essential oils have unique effects on body chemistry along with the ability to alter mood states and promote healing for a wide variety of afflictions. An aromatherapist will listen to your complaints and then prescribe a mix of different essential oils tailored to meet your needs, to be taken as inhalants, to be used in teas or compresses, to self-scent, or to be applied through massage.

In France, and Europe in general, one can receive official certification as an aromatherapist, and homeopathic plant- and herb-based medicaments have been common for decades. Widespread acceptance of natural medicine in North America, however, has been more tentative. Similarly, aromatherapy has enjoyed a longer history and greater popularity in Europe.

Aromatherapy did not arrive in the United States until it made its debut in California in the early 1980s, and no legal certification is required for aromatherapists in the United States, though there are thousands of practioners.

A relaxing massage with pleasantly scented oils will no doubt make you feel good, but aromatherapy promotes the belief that aromas can produce effects on moods and behaviors in intrinsic, druglike ways. Forget the Zoloft and Valium; here are examples of the purported psychological benefits for several commonly "prescribed" essential oils. Sandalwood is sedating and relaxing and is beneficial for treating anxiety, depression, and insomnia. Rosemary clears the mind and stimulates memory. Lavender is uplifting, soothing, and helpful for reducing stress, anxiety, depression, and insomnia. Peppermint is stimulating and strengthening. Sweet marjoram is calming and sedating and helpful in relieving a variety of negative emotional states, including anxiety, irritability, and loneliness. Clary sage is both uplifting and relaxing as well as helpful in relieving depression, anxiety, and fatigue and in calming irritable children. These plant essences, as well as most others used in aromatherapy, are also claimed to relieve a wide range of physical complaints when ingested or applied directly to the skin, including migraine, general pain, gastrointestinal disorders, respiratory problems, and gynecological conditions.*

---

*Note that in contrast to inhaling aromas, ingestion of various plant extracts could have physiological effects, as the natural homeopathy section of your local drugstore attests.

The range of benefits for many of the essential oils in the aromatherapist's coffer sounds unfortunately reminiscent of a carnival charlatan's cure-all. Although many plants do have physiological consequences when ingested, there is no scientific evidence in humans that by inhaling sandalwood aroma the essence of sandalwood is detectable in the bloodstream—which it would have to be if it were producing a pharmacological effect. While lavender is not in danger of becoming a controlled substance anytime soon, lavender does seem to help many people relax. If it isn't working like a drug, then how does it work? The answer is, through psychology.

Aromas work their therapeutic magic by evoking a learned association in the smeller. This learned association can have real emotional and physical consequences, which in turn will influence moods, thoughts, behaviors, and general well-being. The scent of lavender encourages relaxation, and sniffing peppermint is revitalizing because of the meanings these aromas have acquired and the emotional associations they induce. We have "learned" that the emotional association to lavender is relaxation in the same way that we have learned that rose smells *good* and skunk smells *bad*.

The context in which we typically encounter the fragrances common to aromatherapy also helps support their therapeutic effects. Lavender is often found in bath oils and soaps, and since people frequently take baths to relax, lavender is easily construed as relaxing. In contrast, with the help of advertising, its candy context, and the trigeminal cooling produced by menthol, we have linked the concepts of "refreshing" and "stimulating" with mint, and for many people, peppermint is indeed

invigorating. However, as the wintergreen example earlier il-
lustrated,* cultural learning can be divisive in eliciting specific
emotional responses to fragrances. Aromas alter our mood and
calm or excite us due to the emotional associations we have
previously made to them, not because of any inherent or innate
druglike influences they have on us. The joys of aromatherapy
are in the mind of the smeller, produced not by direct action of
an aroma, but by the second-order associations and emotions
the individual has acquired to the fragrance in question.

This is not to say that the effects of fragrances aren't just
as real as they would be if produced by "internal" agents, like
drugs. Studies by Johann Lehrner and his colleagues at Medical University of Vienna have shown that exposure to either
lavender or orange aromas improved mood and reduced anxiety
among patients who were waiting at a dentist's office compared
with a group of patients who were not exposed to aromas while
waiting.[1] The positive therapeutic effects of aroma are not limited to mood, either; they may also influence physical states. In
a study investigating the effects of aroma on pain, participants
inhaled either a self-selected pleasant or unpleasant aroma while
being exposed to painful heat.[2] Those who smelled their "pleasant" odor reported their pain as less uncomfortable than those
who smelled their "unpleasant" odor, even though everyone was
exposed to the same amount of heat.

---

*Comparison of wintergreen preference in the United Kingdom versus the
United States showed vastly different pleasantness perception and could be
traced to cultural associations and the different learned context of wintergreen aroma. See Chapter 2.

These findings highlight another central factor in aroma-therapeutic outcomes. Aromas people *like* elicit pleasant moods and have positive effects, while aromas people *dislike* tend to induce unpleasant moods and have negative or neutral effects. The patients in the dentist office who did not like the scents of orange or lavender would not have benefited from their presence. Illustrating this point, another study on pain perception testing the effects of smelling lavender or rosemary found that neither aroma had any analgesic effect, but people who smelled lavender and liked it rated their pain as lower and their mood as better than those who didn't like the aroma they were smelling.[3] There is a basic logic to this finding: Being in a good mood decreases the intensity of environmental annoyances while being in a bad mood exacerbates them. If you have just had a fight with your spouse, the traffic jam on your way to work is a lot more irritating than if you just returned from a relaxing vacation. An odor you enjoy will make you *feel* good and can decrease your anxiety and increase your tolerance to environmental annoyances, like pain.

SUGGESTIBILITY AND AROMA ILLUSIONS

Our psychological susceptibilities can also fuel the power of aromas through mere suggestion. A clever study by Estelle Campenni at Marywood University showed that simply being *told* that a given fragrance was "relaxing" or "stimulating" produced changes in heart rate consistent with being relaxed or stimulated, regardless of what the fragrance was or even if a fragrance was actually present. Students who were told that

Nose Muzak

"Nose Muzak," as denoted by Reuter's correspondent Sebastian Moffett, is the practice of pumping pleasant fragrances into factories to improve workers' performance. Various Japanese companies have tried this in recent years. The problem is that old devil of odor adaptation. From a psychological perspective, once the new scent, like new furnishings and lighting décor, becomes familiar, it will be taken for granted and no longer perk up employees. From a physical perspective, once aroma adaptation sets in and workers no longer *smell* or notice the scent, it will no longer boost performance. There are no subconscious effects of odors—you need not fear a future of covert odor mind control. Odors cannot manipulate you if you cannot smell them.

lavender scent was "stimulating" experienced a rise in heart rate; those told it was "relaxing" experienced a drop. Importantly, the students were never told what the fragrance was, just that it had the property of being *relaxing* or *stimulating*. Even when there was no fragrance present at all, those given the suggestion that a fragrance in the room "that you may not be able to smell" is "relaxing" showed a decrease in heart rate, and those told that the phantom fragrance was "stimulating" showed an increase.[4] That "suggestion" alone could produce the indicated effects with very popular and stereotyped aromatherapeutic scents like lavender is a serious death knell to the claim that fragrances possess *intrinsic* psychophysiological properties.

It may seem amazing that aromas can be created out of thin air and that these aromas will "affect" us, that we will believe what someone tells us about a scent even when there's nothing there to smell. This is certainly not the case with our other senses. If someone told you that there was a white elephant in the room, you wouldn't suddenly be able to see it. And remarkably, phantom aromas are surprisingly easy to induce.

The first recorded demonstration of an aromatic apparition was made in 1899 by Emory Edmund Slosson, a professor of chemistry at the University of Wyoming.[5] During a college lecture he announced that he wanted to see how rapidly an odor could be diffused through the air and requested that the students raise their hands as soon as they could smell it. He then poured distilled water over cotton, explaining that it was a chemical with a strange odor that nobody would have smelled before. Slosson claimed that within fifteen seconds most of the front row had raised their hands, and within one minute three-quarters of the class had their hands up. In 1978, Michael O'Mahoney, now a professor of food sciences at the University of California, Davis, conducted a very impressive olfactory illusion on British television and radio. O'Mahoney had it announced that a certain sound frequency would produce the perception of an outdoorsy smell. From this mere suggestion and some audiovisual prompts, hundreds of people wrote into the TV and radio station claiming that they smelled something when the frequency was broadcast. Specific scents varied from honey to manure—and in some cases people reported feeling dizzy, getting a coughing attack, and even experiencing a bout of hay fever![6]

This media stunt shows how simply being told that an odor is "there" can convince you that you are smelling something with all of its full-blown consequences. Moreover, when people are left to their own devices to come up with what an imaginary aroma might be and hence do, aromas can be the cause of extreme reactions and hysterics.

In New York City at the end of 2005, several aroma-induced incidents took place that almost certainly would not have occurred without the paranoia instilled by 9/11. In late October and then again in early December, New York City was besieged by a mysterious aroma described by many as smelling like "maple syrup." On October 27, 2005, the sweet, maple, caramel aroma wafted from lower Manhattan to uptown and then into parts of the other boroughs so quickly that fear of chemical terrorism cloaked in the benign disguise of dessert spread from gossip to serious worry. The city's 311 information hotline rang off the hook for twenty-four hours, and the city's Office of Emergency Management alerted police and fire departments, state emergency agencies, and even the U.S. Coast Guard to establish how far reaching the elusive odor was and from where it emanated. Intensive security testing of the air above Manhattan failed to uncover any hazardous chemicals, and within a day the aroma had vanished as inexplicably as it had materialized. A nearly identical maple syrup invasion spread through the Big Apple in early December 2005, and again no source was ever discovered. Although some New Yorkers reported that the aroma revived nostalgic memories from childhood or propelled them to eat sweets they otherwise would resist, the predominant emotional reaction to this scent of maple syrup was dread. In a city haunted by terrorism, a context for an otherwise pleasant aroma was created—danger. The bottom line is that the interpretation and hence experience of a scent is in the mind of the smeller. Fear, illness, happiness, and craving can be conjured from the very same aroma in the very same person.

IS THAT SMELL MAKING YOU SICK?

In the summer of 1999, forty-two children in the small Flemish town of Bornem became mysteriously ill and had to be hospitalized after drinking Coca-Cola. Two days later, eight more children fell ill in the nearby town of Brugge, followed by a spate of similar illnesses across the country. By the end of the week, more than one hundred children were hospitalized with symptoms including nausea, dizziness, and headache, even though most of the sick children didn't drink any Coca-Cola on the day they fell "ill." After a thorough and expensive investigation that resulted in the biggest product recall in Coca-Cola's history, it was discovered that the Coca-Cola production plant in Antwerp had used contaminated carbon dioxide to fizz a batch of fountain drinks delivered to some schools. To Coca-Cola's relief, however, the investigation showed that the level of sulfur compounds in the recalled soda was thousands of times less than the amount necessary to have any health consequences. But it was at the right amount to produce a rotten-egg scent. It was the *odor* that the contaminated soda contained that made people believe they were sick.

Belief in the healing power of aromas is paralleled by a long-standing mythology that aromas can make us sick. There is a chemical factory on the outskirts of Frankfurt, Germany, that emits, among other effluents, a chemical called *thalium*. When thalium comes into contact with human skin, it changes to dimethyl thalium, which at very low concentrations smells like garlic. Whenever the factory neighbors get a whiff of garlic, they rush to their phones to call the plant and complain of the terrible, noxious chemicals that must be afflicting them. However, this factory is also situated next to a forest where a wild plant named barlauch grows, and during the spring, barlauch smells like garlic. Depending on the season and direction of the prevailing wind, the scent of garlic upsetting the local noses could come either from the natural emanations of the forest or from the factory. But no matter, in the springtime the factory constantly gets complaint calls about its malodorous chemical production even when no thalium is being produced. Fear of the miasma is an adaptive survival strategy, but it is often taken to irrational extremes where odors are concerned. "Foul" odors are always guilty first, and we are reluctant to having them proved innocent.

For centuries unpleasant odors have been viewed as carriers of disease while good odors have been believed to be curative. What these "good" and "bad" odors were depended on the time and place. However, throughout history odors that are culturally classified as good are believed to be healthful (in contemporary Western culture, floral, fruity, and minty aromas are some examples), while scents that are culturally disliked— smell "bad" (sulphurous aromas are a current example in North

America)—are perceived as harmful. It is ironic that poisonous household products like detergents and cleaners are often disguised with "good" and "healthful" aromas such as mint, pine, and citrus.

Pam Dalton, of the Monell Chemical Senses Center, has spent a number of years studying how people's personal impressions of an aroma's healthy or harmful qualities go a long way to producing symptoms of wellness or illness. In one study, participants were asked to smell three physically safe odors: wintergreen, whose smell has a healthy connotation; butanol (a version of rubbing alcohol), which easily lends itself to an unsafe connotation; and an unfamiliar woody balsam scent. Pam then did what psychologists often do and lied to her subjects. She told one group that *all* the odors they were smelling were "healthy," another group that the same three odors were *all* "harmful," and a third group that the three odors were *all* "neutral" for health. Here's how this information swayed the participant responses to the scents.

The participants who were told that all the odors were "harmful" reported many more health symptoms, such as nose irritation, sore throat, headache, and dizziness, and gave more intense odor and general irritation ratings to all three odors than the groups who were told that the aromas had a neutral or healthy connotation. It also didn't matter what the scent was. The number of health complaints for wintergreen, balsam, or butanol varied equivalently in accord with being given a "healthy" or "harmful" connotation.[7] In other words, the participant's physical symptoms were produced by words not aromas. The bottom line is that *belief* in the dangerous or bad properties of odors, rather than the reality, is responsible for many health complaints—and can have profound impact on medical policy, economics, and politics.

A DISEASE WITHOUT A CAUSE: MULTIPLE CHEMICAL SENSITIVITIES SYNDROME

Unidentified Foreign Odors—real UFOs—are responsible for more than 50 percent of the reported outbreaks of psychogenic illnesses, controversial illnesses that do not seem to have a discernible physical basis, such as Gulf War syndrome, multiple chemical sensitivities, and sick-building syndrome. A range of odor-induced health complaints without medically visible signs are currently classified under the umbrella of *multiple chemical sensitivities syndrome*. Multiple chemical sensitivities, or "MCS," is a poorly understood and controversial disorder where severe allergic reactions and emotional distress are elicited in the presence of ambient odors. MCS is such a controversial syndrome that there has been a worldwide debate for years about whether it should be classified as a disease or not. In 1992 the American Medical Association stated that MCS should *not* be considered a disease, and similar government organizations such as the Environmental Protection Agency have made similar statements over the years. However, experts, patients, insurance companies, and political institutions have variously taken different views.

The most common symptoms of MCS include fatigue, difficulty concentrating, headache, dizziness, weakness, pounding heart, shortness of breath, anxiety, and muscle and joint tension or pain. The critical piece to this varied list of ailments is that these symptoms are brought on when the person is in the presence of an odor. The aromas that produce these symptoms in any given individual are completely random and can range from grapefruit to perfume to paint. There is a striking range of ailments from physical (e.g., joint pain) to mental difficulties (e.g., concentration problems) and psychological states (e.g., anxiety) associated with MCS. But one common thread is that they are almost always scents that a non-MCS sufferer finds benign. The intensity of an aroma that causes an attack for a particular person can also be as soft as a whisper or as strong as a blast. People with MCS even experience attacks when there are no "smellable" chemicals in their milieu. The only necessary criterion is that the person *believes* the offending odor is there.[8]

On the surface there is no rhyme or reason why a particular fragrance causes an attack in one person, nor is there any consistency among individuals as to what aromas are toxic, except that the offending aromas are often perfume and food related. MCS responses also typically extend beyond odors to include other sensory stimuli, such as noises, lights, and certain tastes. The name "multiple chemical sensitivities" refers to the fact that people who are plagued by this illness seem to be especially sensitive to sensory stimuli of all kinds.

The most striking and controversial feature of MCS is that when people report their symptoms to their physicians, almost invariably no underlying physiological cause can be found. Not only is the MCS patient physically healthy, but in one large-scale study of 264 MCS-afflicted individuals, fewer than 2 percent had symptoms that were in any way traceable to a history of exposure to noxious chemicals.[9] If there are no physical signs that a person has MCS, and in many cases no objective cause is known, how does MCS develop and why does it only afflict some people?

When individuals with MCS describe their illness history, they usually say that there was an original incident involving an offending odor that triggered the onset of attacks, and now, whenever exposed to that scent they become ill. They then usually go on to say that since the onset of their MCS the list of attack-triggering odors has been steadily increasing. What is strange is that among the battery of odors to which a person may claim negative reactivity are many that are far removed from the fragrance that was originally offensive. Here is a classic pattern: Mr. X first experienced an MCS attack—breathlessness, racing heartbeat, and dizziness—in the presence of Giorgio perfume. Like the Belgian schoolchildren who smelled something sulfurous in their Coca-Cola, and the people in Frankfurt who blamed the prevailing garlic aroma as the *reason* they felt unwell, Mr. X attributed to Giorgio perfume the cause of his malady. For Mr. X, however, in future episodes, not only does the scent of Giorgio trigger these symptoms, but the smells of steak, apples, and dish soap do so as well. Why should such a disparate array of scents cause Mr. X to feel sick, when only one was related to feeling ill in the first place? The explanation that I and some other researchers of this topic champion is that steak, apples, and dish soap have become associated with the Giorgio perfume incident such that there develops an ever-increasing spiral of offensive associated scents.

The following scenario illustrates the steps behind the development of MCS in accord with this theory. Imagine that Mr. X's initial attack at a steak restaurant, in the presence of Giorgio perfume, was really a panic attack triggered by some other un-discovered cue. Giorgio perfume, however, is the most distinc-tive and salient cue in Mr. X's environment and because it is a chemical, it is "potentially hazardous." Because the aroma of steak was also in the context of Giorgio perfume, the smell of steak becomes connected and conditioned to his negative reac-tion to the perfume, and the next time Mr. X eats a steak he has a panic/MCS attack. During this second attack, suppose that Mr. X was also having applesauce with his steak, thus a few days later when about to eat an apple he finds himself short of breath—apple scent is now a trigger as well. Eating the apple, however, occurred just after he finished washing dishes; now dish soap fragrance is connected in this chain, and on it goes. Giorgio, steak, apple, and dish soap aromas all trigger MCS symptoms because of their sequentially connected associative history with the original panic "allergic" reaction. The link of odors down this chain can become endless. Indeed some people with MCS complain that they are so disabled by their illness that any noticeable odor (to them) will trigger an outbreak.

According to one medical statistic, between 15 and 30 per-cent of the U.S. population has at one time complained of MCS-like symptoms. Depending on the individual, the condition can be merely annoying or completely disabling. In those disabled by their condition this translates not only to diminished quality of life for the person, but also to lost productivity, increased unem-ployment, and insurance expenses, as well as being an elevated burden on medical resources. In one study of thirty-five patients with occupationally related MCS, 97 percent of the patients had altered their behavior outside of the home, in most cases drastically; 91 percent stated that they now limited their travel; 89 percent decreased their contact with friends; and 77 percent had left a job.[10] Given the considerable social and financial repercussions of MCS there has been a major push by governments to investigate the underlying causes and possible treatments for those afflicted. In 1997, a report by the World Health Organization (WHO) concluded that the need to understand whether the basis for MCS was psychological or toxicological was of "utmost importance" as it would "influence public policy and clinical practice in the prevention and treatment of these complaints."

HOW? WHY? WHAT TO DO?

Despite numerous studies since the 1997 WHO appeal, no clear physiological or toxicological basis for MCS has been discovered. As a consequence of the lack of physical or chemical evidence for MCS complaints, scientists have turned to psychiatric and psychological explanations. And as it happens, about 75 percent of people who report MCS symptoms also have emotional disturbances, typically depression or anxiety.

One psychological explanation for MCS is that it is another form of what is known as a *somatoform disorder,* a condition in which individuals describe a variety of physical complaints, perceive themselves to have a poor quality of life and functioning ability, are continuously seeking help for their problems, and have a tendency to question doctors' opinions. According to this explanation, MCS is just an odor-induced psychogenic illness. However, at least 25 percent of those who complain of MCS symptoms have no prior history of psychiatric diagnosis. Therefore, although psychiatric factors may explain a number of cases, this does not explain them all. So the beguiling question remains: *how* does MCS develop when there is no psychiatric explanation?

The theory mentioned earlier suggests that MCS develops through conditioning with the Rosetta stone of our sense of smell—emotion—at its root. Conditioning, à la Pavlov, will result when an originally neutral stimulus takes on the properties of something "stimulating" simply by being paired simultaneously with it. Meat makes hungry dogs salivate, and if you pair a bell (neutral) with meat (stimulating), over time the sound of the bell will itself take on the stimulating properties of meat and make a hungry dog salivate. We can do the same thing with odors when emotion is involved. That is, we can pair an initially neutral, meaningless odor with an emotional event such that the odor subsequently becomes a proxy for that emotional event and is able to elicit the same feelings, thoughts, and behaviors as the event originally did. Mr. X associated feelings of panic with Giorgio perfume and subsequent arbitrary scents, and then these scents became capable of eliciting the gamut of emotional and physiological symptoms of panic itself.*

MY LABORATORY HAS TESTED THE IDEA that odors can become conditioned to emotions and then act as stand-ins for the emotions themselves, consequently altering behavior in

---

*A psychological basis for MCS does not make the symptoms any less real for the individuals suffering from it than if a toxicological explanation were validated.

accord with the conditioned emotion.[11] In one study we asked five-year-olds to try to complete an "impossible maze" while they were in a room scented with an unfamiliar smell. The maze involved trying to move a toy troll around concentric rings to get to the center of the maze without crossing a line, but the maze was engineered such that this was impossible to do. The children worked on the maze for five minutes and from videotaping their facial expressions and remarks we saw that they became very frustrated by their inability to get to the maze's center. After a short break the children were then brought into a different room and given a new task. The new task was to find and circle drawings of "puppies missing their tails" from a sheet containing 120 animal illustrations.* The key manipulation was that the room where they did the puppy-finding test was either scented with the same smell as the room where they did the maze, scented with a different smell, or not scented at all. We found that children who did the puppy-finding test in a room scented with the same smell as the maze circled far fewer puppies correctly than kids in any other condition. We changed the odors around and got the same results. No matter what the ambient aroma was, if it was the same one that had been present during the impossible maze, the children didn't do well. The children who were exposed to a scent that was associated with a frustrating experience behaved in a frustrated and unmotivated way when later exposed to that aroma.

We repeated a version of this experiment with college students and got the same results.[12] The frustrating emotional experience for the college students was to play a computer game

---

*There were twenty puppies missing tails, and twenty puppies with tails.

that was rigged to be very annoying and make them "lose" in the end, and the room they did this in was scented with an unfamiliar aroma. After a short break the students then entered another room that was either scented with the same smell as the annoying computer-game room, a different smell, or unscented; here they had to work on a series of fairly difficult word puzzles. As a measure of odor-emotional conditioning, we recorded how long the students persisted at solving the word puzzles before giving up. An example problem was "log rail"— *change the order of the letters to make one word.* * We found that students who did the word puzzles in a room that was scented with the same aroma as in the computer-game room gave up much more quickly than the students who did the word puzzles in the presence of a different aroma or no aroma. More specifically, they spent less time on the problems that they ended up skipping and leaving blank than students in the other groups. When confronted with a particularly challenging word problem, the students exposed to the computer-game-room aroma gave up far more quickly than the other students. The emotions of frustration and annoyance had become associated with the ambient scent, which consequently made them feel and behave in an unmotivated way.

A large body of psychological evidence demonstrates that how we feel influences how we think and how we act. When we feel happy, we behave happily as well: helping others, paying more attention, being more creative, and so on. When we feel bad, we generally behave in less productive and more antisocial

---

*The answer is "gorilla."

ways than normal. Our olfactory system is wired to readily form connections with meaningful information. It is also in this way that emotions become attached to odors and why an odor can elicit the same emotions as an original situation did.

So how does this explain MCS? For someone who experiences MCS symptoms, there was an original situation where the person experienced strong negative emotions, such as anxiety, along with physical consequences, such as hyperventilating, while a distinctive scent was also present. Hyperventilation changes the pH of one's blood because of the imbalanced exchange between carbon dioxide and oxygen and leads to a host of other physical symptoms, including shortness of breath, rapid heartbeat, dizziness, numbness, and supreme anxiety or panic. The physical and emotional symptoms induced by hyperventilating then become attached to the aroma that was present, and subsequently whenever the odor is present it can itself induce the physical symptoms of actual hyperventilation and feelings of anxiety or panic.

To test this idea, Omer Van den Bergh and his colleagues at the University of Leuven in Belgium had participants inhale carbon dioxide, which causes hyperventilation, while an odor was in the air and later examined breathing reactions to the odor by itself. The researchers found that persons who had undergone the carbon dioxide–induced hyperventilating association started to hyperventilate when they were exposed to the scent; and these reactions continued to occur in response to that odor for at least a week after the association was first made. Interestingly, several different odors were tested but not all of them could be conditioned to cause hyperventilation.

Only odors that the participants didn't like, such as ammonia or butyric acid, could be linked to fear and physical symptoms of panic; floral odors, such as neroli, that were classified as neutral or pleasant, did not become conditioned to the panic symptoms. Another interesting finding was that participants who were classified as psychologically "neurotic" were much more susceptible to the hyperventilation manipulation than nonneurotic individuals. That is, participants with **negative** feelings about the world, who generally noticed their **bodily sensations**, and were likely to worry about their health, a personality profile common in MCS, were particularly inclined to connect an adverse physical response with an aroma.[13]

While there is substantial evidence that MCS is a psychological condition, public policy has been much more aimed at attacking the supposed "toxicology" of odors than dealing with underlying psychological issues. In California, a number of towns and communities sport notices in public places against the wearing of chemicals (perfume), and many other locations around the country have instituted similar policies. Nowhere, however, is the hazardous potential of aromas taken so seriously as in Halifax, Nova Scotia, where it is actually *illegal* to wear fragrance in public places. Application of this law has been taken to such absurd extremes that two incidents made national Canadian news. In one case, an emergency room nurse who had just arrived for her shift freshly showered and powdered was told that she had to go home and take another shower before returning to work because the fragrances or products she had used could be smelled. In another case, an elderly woman was refused entry onto a public bus because she was wearing perfume. While it is possible that some odors may be toxic to some people, this condition is rare compared with the number of individuals who have learned negative associations to odors. Nevertheless, governments seem to find it easier to deal with aromas than to deal with psychology.

AROMA AND PSYCHOLOGICAL THERAPY

There is no doubt that our sense of smell is more intensely and intimately linked to our moods and emotional life than any of our other sensory experiences. Therefore, one would expect that fragrances could also be effectively used in genuine psychological therapy. Indeed, some well-respected clinicians have put aromas to use on the couch. One of the primary ways in which odors have been used in psychological treatment is in helping people overcome anxiety. For example, a fragrance associated to feeling calm can be used to help anxious people relax. One aroma with clinical success in England is "maratima" (scent of the sea). Note, however, this aroma would only be successful among those who had positive and calming associations with sea scent, a good bet in the United Kingdom. But if you had negative associations to sea scent, it would not be therapeutic. It could even be a trigger of a PTSD attack for survivors of a tsunami. The associations that occur between emotions and odors are based on our personal experiences and may not be predictable; thus, the pleasure or pain that odors elicit is not the same for all of us.

A discussion of psychology and therapy is not complete without mentioning Sigmund Freud, who at one time believed that our sense of smell, in particular the nose, was profoundly connected to our sexuality and to the hidden psychological problems manifested by his patients. To summarize briefly, and admittedly without justice, "neuroses" were the dysfunctional behaviors and emotional problems that Freud's patients exhibited. Freud believed that these neuroses were due to his patients' subconsciously repressed conflicts and memories from childhood that were primarily sexual in nature. The task of psychoanalysis, Freud's method of therapy, was to find the repressions that were causing the neurosis and to bring them to consciousness where they could be dealt with and defused. In some cases, however, physical interventions were deemed necessary to bring these repressions to the fore. Enter Wilhelm Fleiss. Fleiss, a German otolaryngologist, became close friends with Freud in his early career. Fleiss was obsessed with the notion that the nose and the genitals (sexuality) were connected and that dysfunction in this connection was a major reason why Freud's patients were neurotic. This notion, coupled with his medical training, led Fleiss to conclude that Freud's patients could be cured by nasal surgery. Accordingly, Freud sometimes referred his patients to Fleiss for treatment of their neuroses by anesthetization of the nasal passages with cocaine, a drug that both Freud and Fleiss were indulging in at that time. Freud's connection with Fleiss and the nasal treatment of neuroses, however, came to a screeching halt with the disastrous surgery of Emma Eckstein.

At the age of twenty-seven, five years after Fleiss and Freud had become close associates, Ms. Eckstein came to see Freud with complaints of vague symptoms, including stomachache and mild depression related to her menstrual cycle. Today Emma would have been diagnosed with PMS, or premenstrual syndrome. Freud, however, diagnosed Emma as suffering from a "nasal reflex neurosis" and referred her to Wilhelm Fleiss. Fleiss removed Emma's turbinate bones, the bones that give our noses their shape and structure. The surgery was a disaster and infection quickly broke out. A second surgeon was called in and found surgical gauze that Fleiss had left behind in Emma's nose. Its abrupt removal caused a hemorrhage so severe that Emma nearly bled to death. Emma survived but was rendered permanently disfigured and anosmic. Shockingly, Emma remained on good terms with Freud and became a psychoanalyst herself. Freud also didn't sever his friendship with Fleiss until much later, due to accusations of plagiarism, at which point he ordered that all his correspondences with Fleiss be destroyed. Marie Bonaparte, Napoleon's great-grandniece and a psychoanalyst herself, bought their letters and prevented their destruction, which is how we know of this catastrophe today.

In part due to Freud's legacy, our sense of smell has been ordained as our most sexual of senses, and throughout history specific scents have been heralded to possess aphrodisiac powers. As you will see in the next chapter the quest for an aphrodisiac in a bottle has to date been a multi-billion-dollar quest—without success—yet its allure and pursuit remain unabated.

CONDITIONED LUST

A link between sexual fetishes and scent has been noted by other psychologists apart from Freud. Havelock Ellis (1859–1939), a famous English psychologist, described a young woman patient with a fetish for the scent of leather who could experience orgasms in the presence of leather objects, particularly leather-bound ledgers, and in leather shops. Ellis traced the origin of this fetish back to the woman's memory of an early experience masturbating in the presence of a distinctive leather scent. Due to her past experiences connecting orgasm with the scent of leather, her orgasms had become conditioned to leather aroma; therefore, whenever this woman smelled leather she would spontaneously experience full-fledged sexual climax. This is an intense example of how scents can hijack our minds and our bodies through emotional associations.

CHAPTER 5

SCENTS AND SENSUALITY

I knew I was going to marry my husband the
minute I first smelled him.

—ESTELLE CAMPENNI, 1995
(psychology professor, Marywood University)

We were having coffee, talking about the usual things women
talk about when we are first becoming friends, especially when
one finds out the other has a special knowledge about an
enigmatic and intimate facet of our lives. Estelle, I quickly
learned, was immersed in her sense of smell to such a degree
that she trusted her nose to make life-altering decisions. "I knew
I would marry my husband the minute I first smelled him" she
confided. "I've always been into smell, but this was different, he
really smelled *good* to me. . . . His scent made me feel safe and
at the same time turned on . . . and I'm talking about his real
body smell, not cologne or soap. I'd never felt like that from a
man's smell before, and it was so compelling I knew he had to
be right for me. And it's true. We've been married for eight years
now and have three kids, and it's still great and his smell is

always very sexy to me. Funny, isn't it? Before Harry, I'd often been turned off by how a guy smelled, and it wasn't because he hadn't showered." "Me, too," I replied. "I didn't really think it had much significance," Estelle continued. "Does it? What does it mean?"

It means that *body chemistry* plays a startlingly large role in who we are sexually attracted to and that our noses speak loudly to our souls even if it seems like only a barely audible whisper.

Since at least the time of the Egyptians, we have been adorning ourselves with scent for the purposes of attraction. However, we needn't be bothering so much with daubing the essences of spices, flowers, and plants on our bodies, because our own natural aroma is far more powerful for stimulating attraction to one another than the bouquets of the garden. Many cultures have recognized the special relationship between smell, affection, and sexuality. In India the word for "kiss" means smell, and the Inuit typically rub their noses together rather than engaging in mouth-to-mouth kissing. Members of a tribe in New Guinea say good-bye to each other by putting a hand in each other's armpit and then stroking it over themselves. And there is the oft-portrayed scene from the Elizabethan period, of paramours exchanging "love apples," where a woman would keep a peeled apple in her armpit until it was saturated with her sweat and then give it to her sweetheart as a scent token. Coco Chanel said, "Without perfume women have no future," but I think she was wrong. Our attraction to each other's "real smell" is vital to our sexuality and reproductive success.

BODY CHEMISTRY

Most people don't trust their noses nearly as much as Estelle does, and they would never admit to making life-altering decisions based on a scent. But all of us, and women in particular, end up relying quite deeply on our noses for some of the most fundamental decisions in our lives; in particular, whom we choose to have children with. The sense of a lover being "right" or "wrong" for you may feel like intuition, but it is real—and your nose knows it.

There is something very important and biologically sound going on when a woman finds a man's smell delicious or distasteful. Men can also tune in to the important scent messages in a woman's bouquet, but they tend to rely more on their eyes than their noses when it comes to sexual attraction. This isn't about pheromones, which I'll get to later in this chapter. This is about real body chemistry, and the basis for its significance is in evolutionary biology.

In the tenets of evolutionary theory the ultimate goal of each individual is to ensure that his or her genetic makeup is represented as much and as widely as possible in future generations. In other words, *go forth and multiply.* In biological terms, the best outcome for you is that your traits (genes) will be more prevalent in current and future generations than your neighbor's traits. This is "survival of the fittest," and the drive to be "fit" underlies our fundamental motivation to have children. But men and women are not created equal and as such have different costs and benefits to lose or gain by reproducing and parenting. As a consequence, nature has equipped men and women with different interests and goals in playing the game of love.*

Women, like other mammals, "invest" far more than men do when it comes to having a child. In addition to nine months of pregnancy, which requires about a 40 percent increase in energy expenditure, there is also a period of at least one year postdelivery where, before the invention of milk formula, a mother could not become pregnant again or she would stop lactating and her infant would starve. Another limitation is that a woman can only become pregnant from one man at a time. Therefore, a woman has to invest at least two years of her life and significant physical and psychological resources to ensure that the child from only one man will be a healthy baby and then at least another thirteen years or so of care to ensure that the child is reproductively successful him- or herself. These are heavy costs for a woman to bear and yet her biological success is measured just as a man's is—by how many healthy and reproductively successful children she can produce. Although the world record for the number of children born to one woman is sixty-nine—most of them being triplets—ten births is a realistic maximum.[1]

In contrast to women, men have almost nonexistent physical costs, needing only minutes to engage in the act of sex. And because men have millions of continuously replenishing sperm, they can, and some do, have a multitude of different sexual

---

*Evolutionary principles are concerned only with gene propagation and how this influences our behavior and do not consider homosexuality or nonprocreative motivations like becoming a priest.

partners—each with the biological potential of becoming pregnant. Technically the only limiting factor for a man is the number of hours in the day. King Moulay Ismail the Bloodthirsty, a medieval Moroccan despot, is reported to have sired more than eight hundred children by the women in his harem.[2] But there is one big catch—prior to the advent of genetic testing, there was no way for a man, even an egomaniacal one, to be absolutely certain that the child presented to him was actually his. A man must rely on the honesty of the woman who gave birth to the baby to reassure him that he is indeed Daddy. For a woman, however, there is no doubt that her child is her own. Therefore, when a woman invests in her child she knows she is investing in her own genetic material. The challenge for a man is whether he should give resources to a woman who may have cheated on him and support a child that may not be his own. This fundamental difference between men and women, and the evolutionary principles of this primitive conundrum, are the biological roots for many of the worst aspects of human behavior, including jealousy, child abuse, and spousal homicide.[3]

You can witness the primitive impulses behind sexual commitment and the different priorities of Mars and Venus the next time you are in a gathering of male and female friends. If you dare, ask each of them to consider which situation would "hypothetically" make them feel more jealous: (1) your partner/boy(girl)friend/spouse has just confessed to you that s/he had sex with someone else while away on her/his last business trip but that it meant nothing, or (2) your partner/boy(girl)friend/spouse has just confessed to you that s/he has fallen deeply in love with someone else but has not had sex with him/her.

These scenarios were in fact put to men and women in a study conducted by David Buss and his colleagues at the University of Michigan. They found that most men rated situation #1 as worse than #2, and that most women rated situation #2 as worse than #1.[4] I have also informally tested these same scenarios with students in my class and gotten the same results. So if your friends are telling the truth, they are likely to answer similarly. These primitive emotional responses to emotional and sexual fidelity are driven by our core biology and its gender-defined challenges and concerns. Situation #1 reveals the threat of cuckolding and signals that a child you may be investing in with your partner may not be yours, whereas situation #2 reveals the threat that emotional commitment and the resources, protection, and assistance it provides to you and your children is going to be withdrawn.

Because both women and children benefit greatly when mothers get help raising their babies, traditional research on heterosexual attraction has presumed that the most important attributes a woman seeks in a man are signs that he will be a good protector and provider, able to take care of her and ensure her offspring's success. Modern indicators of "provider" tend to be evaluated by status symbols, such as the kind of car a man drives, his job, and the size of his bank account. By contrast, because men can *potentially* impregnate countless women but may in fact impregnate none, the most beneficial strategy for men is to find receptive females who look like they would be able to produce a child. *Looks* is the operative word, as the features universally regarded as physically attractive in women— lustrous hair, clear skin and eyes, and a hip-to-waist ratio of approximately 7:10—are in fact highly correlated with fertility. So to be the most evolutionarily successful and get the most genes out into future generations, men should be biologically programmed to find good-looking, young women, while women should be searching for "sugar daddies." There is copious survey data supporting the theory that women are attracted to rich and powerful men, and that men turn their heads to the young and beautiful. But something about this cut-and-dried dichotomy— men seek high-status physical features and women seek high-status social ones—always struck me as not quite right.

I reasoned that although having a man around who is a good provider will increase the likelihood that a child will thrive and be a parent one day her- or himself, it could not be nearly as important as whether or not the child is healthy in the first place—because nothing is more important to survival and being able to reproduce than being hale and hearty. So what should the first and foremost quality be when a woman hunts for a father for her children? Not driving a Ferrari or being a CEO. Rather, a woman should be most attuned to cues that reveal the quality of a man's innate health.

Not so long ago, a frail and sickly child, especially if born to poor parents, would be unlikely to live until his or her thirteenth birthday. Our immune system is responsible for what diseases we can fend off and how well, and also what diseases we may carry as recessive traits. How well we can battle tuberculosis or if we are born with the misfortune of cystic fibrosis is determined by our immune system and the genes we inherit. Since a woman is trying to maximize the likelihood that her child will survive and be reproductively successful, she should want to find a father who has a healthy immune system and, even more important, an immune system that is complementary to her own.

Having children with someone who has a genetically similar immune system as you doubles your children's chance of receiving recessive traits that you carry, which means your children run the risk of developing illnesses that can kill them before they are able to become reproductively active. Tay-Sachs and cystic fibrosis are examples of what can happen. By contrast, if you mix your genes with someone who has a genetically different immune system from your own, your children will have much greater disease protection and will be unlikely to inherit unpleasant recessive syndromes. In other words, you will have children with a high chance of surviving and reproducing themselves. So the most critical biological goal for a woman under the motivation of her selfish genes should be to find a healthy man who has a different immune system from her own. But how can a woman know what a man's immune system is like?

Every biological feature that you have has both a genetic internal manifestation (like the gene for eye color) called a *genotype,* and an outward physical manifestation of that gene called a *phenotype* (your blue eyes). Our immune systems are coded for by a cluster of genes called the *major histocompatibilty complex,* or *MHC.* The MHC comprises more than fifty closely connected genes strung together along a single chromosome, and MHC genes are the most variable in all of nature. It turns out that everyone, unless you have an identical twin, has a unique set of MHC genes. Your unique string of MHC genes is the genotype for your immune system, and your phenotype, the external manifestation of the genes for your immune system, is your body odor.[5] Your MHC genes determine your unique immune system and also your unique odorprint, and your odorprint is as unique as your fingerprint. This explains how a tracking dog finds a prison escapee by smelling his sock and doesn't make the mistake of tracking down the mail carrier. Only identical twins cannot be distinguished by body odor. Identical twins share 100 percent of their genes with each other, and if they eat the same diet there is no way for even the keenest canine noses to tell them apart.

OF MICE AND MEN

To help us better understand ourselves it is often useful to turn to our animal brethren. Mice who are genetically identical except for minor variations in MHC will preferentially select mates who are dissimilar at these same points of variation. The way mice make this decision is by mouse body odor, and it is the *female* mouse who makes these odor-based selections.

But is it a leap to go from mice to men? No identical twin breeding experiment could ever be ethically sanctioned, but it just so happens that a natural situation allows us a glimpse into a real-life experiment—the North American Hutterites. The Hutterites living in the United States today originated from an Anabaptist religious group from the Tyrolean Alps that formed in 1528. Religious persecution forced this sect to relocate repeatedly throughout Europe, and finally in the 1870s, about four hundred members escaped across the Atlantic and settled in the area that is now South Dakota. The original four hundred then split into three communal farms, or colonies. The founding members of these three colonies are the ancestors to all the Hutterites in the more than three hundred fifty clans today, which now extend through the Dakotas, Minnesota, and western Canada. An important feature of Hutterite life is that by self-decree sect members minimize their interactions with mainstream North American culture and maintain strict traditions on mixed-sex social encounters.*

When a Hutterite girl becomes a teenager she and her of-age sisters typically go to spend their summers at the home of another family usually within their same clan. They also go to visit these families throughout the year to help with chores and other domestic events. The boys from these "visited" families, however, remain at their own homes. Somehow through these mix-and-mingle encounters, future marital pairs among adolescents are formed. Given the close genetic relatedness between individuals within the same clan it comes as a shock that deleterious recessive traits don't run rampant and, rather, that healthy babies are the norm. With the very limited choices in marriage partners, how is "inbreeding" avoided?

Carol Ober, a geneticist at the University of Chicago who has been studying the Hutterites for years, found that Hutterite choices in marriage partners are not random. Far more often than would be predicted by any chance romantic encounter, individuals from related families end up marrying partners who have the most *different* MHC from themselves that the

---

*Younger members today may not be following this edict as strictly as would be desired by the community elders.

group affords. That is, sisters from family A end up with the brothers in family B who are most different from themselves in MHC genes, even when age and other social factors are taken into account. And it seems that the Hutterites are using their noses to do this.[6]

Claus Wedekind, a zoologist, and his colleagues at the University of Bern in Switzerland have taken the MHC body-odor attraction question into the laboratory to test the theory that women choose men as sexual partners on the basis of their body odor and its correlated MHC match.[7] To conduct this test, a group of young men and women were profiled for their MHC genes, and then women were asked to choose what men they found sexiest on the basis of their body odor. The men in this study wore a cotton T-shirt to bed for two consecutive nights, adhered to a strict bathing schedule, and were told to avoid activities that would produce "disturbing smells," such as drinking, eating onions, and having sex. After this forty-eight-hour regimen, the T-shirts were collected and placed in identical cardboard boxes. Each woman then smelled the odor emanating from six boxes and was asked to indicate what box/T-shirt she thought smelled the sexiest and most pleasant. Three of the boxes contained T-shirts from men that were similar to her in MHC genes and three from men who were different.

Wedekind found that the women consistently picked T-shirts as the sexiest and most pleasant smelling from men who were the most different from them in MHC type and, consequently, had the most dissimilar immune systems. Because of the biological benefits of complementarity in immune genetics and not simply a healthy immune system overall, this implies that there is no Brad Pitt of body odor. We all possess different MHC genes (and body odor), so for every woman a different set of men will be delicious smelling, while another set will be unattractive because of their body odor. A man whose scent I find scintillating may be ignored or even disliked by my best friend. It is our *chemistry,* our unique body-odor chemistry, that makes it click between a man and a woman.

Wedekind's findings seem to neatly indicate that the right set of genes and a good shower is all a man needs to sweep his biologically most suitable mate off her feet. But there was another finding in the Wedekind study that muddies the water. It turned out that only women who were *not* taking birth control pills found the body odor from "different" MHC guys the sexiest. Women on the pill had the opposite tendency and picked as best smelling the men who were most genetically similar to them. Why?

The birth control pill hormonally mimics the state of pregnancy, and during pregnancy a woman's vulnerability to threats and danger is heightened. For protection, being around members of your family may make you safer than being around strangers. It is therefore speculated that there is a biological predisposition for pregnant or "hormonally pregnant" women to seek men who, because they are more genetically similar, may also be more protective.

This idea is given further credence by a strange phenomenon observed in rodents, known as the *Bruce effect.* The Bruce effect occurs when a pregnant female gets a whiff of a "strange male," a male who is not the sire of the offspring she is carrying, and then spontaneously aborts her litter. Although this is disastrous for her reproductive success in the near term, it means that she is now immediately available to become pregnant with the new male. Had she not aborted her litter it is very likely that the new stud would cannibalize her young once they were born anyway. The phenomenon of spontaneous abortion when women are in the presence of unrelated men has not been documented, and there are no data to refer to. Nevertheless, it is tempting to wonder whether the tradition of sequestering women from social society while they are pregnant, which is practiced by some religious groups, has its basis in the possibility that miscarriages are higher when pregnant women keep company with many unrelated men. Cannibalism by stepfathers is also certainly not within the normal human repertoire. However, child abuse is perpetrated with greater frequency by nonblood fathers than by blood fathers.[8]

HOW DOES IT WORK?

How does this scent sensing–genetic recognition system work? Is some ancient part of our brain subconsciously pointing us toward genetically compatible mates? Or is there something more psychological about why women prefer men whose body odor signifies a compatible genetic match? While these questions cannot yet be fully explained, there is evidence that both biology and psychology play a role.

It so happens that a woman's sense of smell is better than a man's, but only during those few days within a woman's monthly cycle when conception is possible. During the rest of her cycle she is no better than a man at detecting scents, and during menstruation, her ability to detect odors is *worse* than a man's. It is no accident that a woman's greatest nasal acuity coincides with the fleeting days of maximum fertility—the time when it is most important for her to sense out the best biological mate.

Biological determinism, however, can also be turned on its head. You saw earlier that mice select genetically dissimilar mice as mates when given the choice. But a clever study by Dustin Penn and Wayne Potts at the University of Florida put the idea that there is an innate biological sensor behind this behavior into serious question.[9] Penn and Potts found that when mice are fostered and raised in genetically dissimilar litters and later given the choice to mate with (1) their real genetic siblings who they *were not* raised with or (2) genetically dissimilar mice who they *were* raised with, they will choose their real genetic siblings as mates. That is, they make a genetic MHC mistake in mate choice, but in doing so, they avoid mating with a familiar-smelling mouse. In other words, the mate-choice response is based on experience—psychology not biology. Could avoidance of the familiar also be behind our preferences for genetically different-smelling mates? Do women simply avoid men who smell like "family" whether they are actually related or not? There have been no studies in humans that can directly test this, but it would be fascinating to know how this relates to the incest taboo, and whether people who are adopted would be more sexually attracted to the familiar scent of members of their foster family, with different MHC genes, or members of their biological family, whose scent would be less familiar but with whom they would share more MHC similarity.

THE RAINBOW FLAG

The theory of MHC compatibility and its evolutionary consequences for survival is founded on the idea that the primary goal of sexual behavior is reproduction, not pleasure. Therefore nonreproductive intercourse, such as gay and lesbian sex, would not be expected to follow similar rules. But does it? Although little is known about gay and lesbian responses to body odor in relation to sexuality, a recent study conducted at the Monell Chemical Senses Center found that gay men prefer the smell of gay male sweat compared to the smell of sweat from either straight men or women.[10] The participants in this study did not know that the sweat they were smelling was from gay or straight donors or from males or females. This finding suggests that there must be something different in the underarm secretions of homosexuals versus heterosexuals, and that preferences for one or the other is based on something biologically revealing about sexual orientation. Regardless of ultimate evolutionary intent, it seems that for both pleasure and survival of the species, how much you like someone's scent is a crucial driver in sexual attraction.

LIFESTYLES OF THE RICH AND FAMOUS

How do we reconcile the importance of body odor with the tabloid headlines and survey data that young attractive women want the Donald Trumps of this world? Is being rich more important than smelling "right"? To answer this question, my laboratory conducted two large surveys and asked young women how important they considered a variety of physical factors to be when deciding on a man as a potential lover, including looks and body odor, as well as social factors, such as money, intelligence, and ambition. We found that above all other physical characteristics, women ranked a man's scent as the most important feature for determining whether she would be sexually attracted to him. We also discovered that the reason body odor stands out isn't because women are actively avoiding men who smell bad; rather, they were enthusiastically seeking men who smelled gorgeous to them. In the words of one respondent: "If I'm with a guy who smells really good, nothing else about him seems to matter."

This statement also points to a biologically worrisome finding we uncovered with regard to the spell cast by a man's scent. We discovered that women don't differentiate much between real body odor and fragrance. In other words, the Axe commercials are right. If a man smells great to a particular woman, she will find him attractive no matter where the good scent comes from. Indeed, the power of fragrance is such a strong sexual stimulant for women that Dr. Cynthia Graham, while working at the Kinsey Institute,* found that a popular cologne for men was able to enhance women's fantasies and intensify their sexual arousal.

The command of a man's store-bought fragrance is worrisome because it means that a biologically unsuitable man can trick a woman into being with him by "falsely" smelling

---

*The Kinsey Institute at Indiana University promotes interdisciplinary research and scholarship in the fields of human sexuality, gender, and reproduction.

scrumptious. Once a love connection has been forged, a woman is very unlikely to break up with the object of her love when his true body odor is revealed. This is because all the positive emotions surrounding her lover will have become associated with his scent, which then becomes attractive simply because it is associated with him. (I used to be partial to the scent of bicycle grease because my boyfriend at the time worked in a bicycle repair shop.) However, the threat of seduction by scent-disguised men may not be entirely biologically perilous. In a recent study it was found that people's preferences for specific perfumes and colognes was correlated with their own MHC type.[11] That is, men with similar MHC types are likely to pick the same colognes to wear. It may also be that what we choose to wear as fragrance amplifies certain aspects of our own body odor, so colognes may not totally mask a man's scent; they may also mimic or enhance it.

Being in love not only makes the object of your affection more redolent, it may even change your sense of smell. Christie Fowler and her colleagues at Florida State University found that when a male prairie vole looks at a female prairie vole he grows extra neurons in his olfactory cortex. Pair bonding among prairie voles therefore may be based on the sexual imprinting of scent—"you are the scent that I am in love with." Could similar mechanisms be at work in other mammals, even us? Indeed, both men and women report smelling the unwashed clothing of absent lovers to conjure their presence. Thus it seems that we, too, form olfactory love maps.[12] Activation in the amygdala, the emotional nexus of our brain and central axis to our sense of smell, is more intense when we see someone we are in love with compared with when we are looking at a friend. My neuroimaging work also predicts that our amygdala would become more excited by the scent of a lover than the scent of a chum.

But what of rich men? In our survey we did find that women rated money and ambition as more important attributes in men than men did in women. How does body odor fit in with this? Could it be true that being rich makes you smell better?

THE SCENT OF HEALTH

A variety of illnesses are heralded by bad body odor. Before blood tests and other internal measures were available, doctors routinely relied on their noses to diagnose diseases. According to the medical nose, typhus smells like a mouse colony; diabetes, fruit; the plague, soft apples; measles, freshly plucked feathers; yellow fever, the butcher shop; and kidney failure, ammonia. The scent of acetone on someone's breath is also an indicator of diabetes. Certain medications, such as those given to organ transplant recipients and patients with schizophrenia and AIDS, can also cause unpleasant changes in body odor and breath odor. But it isn't only that certain characteristic emanations signify specific illnesses. It turns out that having a generally pleasant body odor is correlated with having generally good physical health. This is because of another physical feature called *symmetry*. Symmetry refers to the size equivalence between body parts that you have two of. You would be highly symmetrical if your right and left ears, your right and left eyes, and your right and left wrists and feet were the exact same size, and less symmetrical if one or more of these left-right pairs didn't match up.

Symmetry isn't just a signal of health, it is also a classic characteristic of attractiveness. Both the same and opposite sex judge more symmetrical faces as more attractive than less symmetrical faces. Of particular interest here is that Randy Thornhill and Karl Grammer while working together at the University of Vienna found that people who were more symmetrical were healthier than less symmetrical people.[13] But what does the size of your left and right foot have to do with how healthy you are?

The explanation is that physical symmetry indicates strong resistance to parasites and other onslaughts during development. A growing fetus might have to endure a variety of physical assaults through its mother's experiences, ranging from viral and bacterial infections to alcohol or drug poisoning, lack of nutrition, or physical abuse. The more robust your immune system is, the more capable your body is at fighting illness and remaining healthy. The theory contends that physical beauty— in particular, symmetry—is a sign that an individual is physically hale and able to withstand hazards that would threaten health. By contrast, unattractive people may be unattractive in part because their immune system wasn't capable of warding off, or didn't do a very good job at defending against, these physical assaults during early development. Though we may find the ones we love *beautiful* no matter what they look like, research has shown that beauty is much less in the eye of the beholder than in consistent physical traits that signify health, strength, and fertility. Therefore, if you are beautiful and symmetrical, it is likely that you are also healthy and are endowed with a tough immune system. What is especially interesting is that women judge symmetrical men as having better-smelling body odor than less symmetrical men. In other words, smelling *good* advertises that you are healthy and attractive in general.

Through many stepping-stones and correlations I have an explanation that is consistent with current research, to account for the connection between being rich and successful and having a pleasant body odor. On average, rich successful men are healthier than unsuccessful men because it is hard to be successful if you are constantly plagued by illness. Moreover, general good health makes men more physically and nasally attractive than their less healthy competitors. This does not necessarily mean that Donald Trump is the right MHC match for model Melania Knauss, but it does mean that he probably smells better fresh out of the shower than someone his same age whom you meet at the unemployment center. Rich men have been dealt a lot of good cards, and this means they can get more tricks.

All this is not to say that men don't find women's scent attractive, too. In our surveys we found that men ranked how a woman smells as second in importance to how she looks and more important than how her voice sounds or how her skin feels. The *Kama Sutra*, the fifth-century Hindu sex manual, written by men, also extols the scent of women and proclaims that the beauty of a woman is not determined by how she looks, but rather how she smells.

Inspired by the Hindu scripture, the late Dr. Mookergee, an Indian chemist at International Flavors and Fragrances, one of the big five aromachemical houses, tried to capture the scent of beautiful women to see what their body odor was like. Using a technique called "Living Flower," where aromatic molecules that are given off by a flower, or a human, can be captured and analyzed, Dr. Mookerjee discovered that the majority of the scents given off by beautiful women were flowerlike, especially lotus and cotton flower. The allusion of beautiful women to flowers may indeed have its basis in reality.

FROM DATING TO DIVORCE

Jessica Ross confided in me that one of the results of her anosmia and one of the main causes for her current emotional distress was that she could no longer smell her husband's scent. Her ability to feel close to him and comforted in his embrace was strongly induced by his smell. "One of the things that attracted me to my husband was his wonderful, sexy smell. Now that I can't smell him I'm not as attracted to him anymore. It also makes me much less interested in being intimate with him. I just can't love him the same way I used to."

Besides having biological significance for healthy children and fertility, the connection between genetics, body odor, and attraction also has profound social implications, and modern life may be messing things up. First there's the "false advertising" men can do by wearing fragrance. And then there is the fact that women who take birth control pills prefer the scent of men who are more genetically similar to them and consequently less biologically compatible. This raises the question of whether the pill confuses a woman's choice in men and orients them toward men who are too genetically similar for successful fertility. Could the misleading scent impressions produced by the pill also eventually lead to marital instability? Archival data from fertility clinics indicates that couples who are more similar in MHC have greater difficulty conceiving. Depending on the dynamics and goals within the couple, fertility problems could lead to tension and make marital waters turbulent. But there is another mechanism by which the pill might also undermine wedded bliss.

> Denise stared off into the middle distance and sighed as she admitted to her psychiatrist: "What can I say, I know the relationship is over and everything he does bothers me. It used to be that when I'd be angry at him and he'd try to win me back and hug me I'd feel all right again in his arms. Now I don't want to be anywhere near him, and hugging him is the worst. I just can't stand how he smells!"

From 1960 to the present day, the birth control pill has been the most common form of contraception in North America. However, it is usually the case that after having children or after a number of years of cohabitation, women switch to nonsteroidal methods of family planning. Could it be that women who meet their husbands while on the pill and then change to nonsteroidal forms of contraception during marriage become less enthralled by their husband's aroma because of a change in their perception of his scent? Is it possible that being on the pill produces subtle changes in a woman's sense of smell such that after going off the pill, some people—like one's husband—no longer smell the same?

More than 50 percent of North American marriages end in divorce. If MHC compatibility is low, a woman may discover that she is no longer nasally attracted to her partner and his changed scent, now unpleasant, can actually be a mediating factor in their breakup. Indeed, I have been told by several marital therapists that one of the most common complaints made by women is a current repulsion for their husband's smell. In light of the converging evidence, when interviewed on the topic of body odor and attraction, I suggest that women who are on the hunt for lifetime partners and the father of their children might want to consider going off the pill before embarking on this quest. However, I also mention that regardless of a man's biological suitability, through "learning" (as previously described), the accumulation of years of negative emotional associations can turn a man's once perfect scent into a signifier for loathing, in which case the pill has nothing to do with falling out of love. If only sexual passion could be kept alive forever or on tap anytime it was desired.

PHEROMONES: FACTS AND FANTASIES

Many people, including mavens in the fragrance industry, believe that aromatic aphrodisiacs exist and are only waiting to be found. The competition in the commercial world for this holy grail is fierce because if such a magical elixir could be captured, it would be the beginning of a trillion-dollar industry, not to mention the solution to loneliness and guaranteed success for nights out on the prowl. Type the word *pheromone* into Google and you will get hundreds of hits for companies offering to sell you pheromones "guaranteed for sexual success." But does such a potion really exist?

One of the most well-known and first among the pheromone merchants is Dr. Winifred Cutler of the Athena Institute. Dr. Cutler sells "Athena pheromone 10X" for men, and "Athena pheromone 10-13" for women, and she has "data" showing that these elixirs boost the sex life of those wearing them. But think for a minute about how we might also interpret data that wearing pheromones makes those donning it more sexually successful. The Wonderbra Company, makers of said bra, used a wonderful truism in a Canadian ad jingle of the early 1970s: "If you look good you feel good, and if you feel good you look great." There are stacks of psychological literature showing how feeling good about yourself makes others perceive you as more attractive. If you believed that putting something on—whatever it might be—would make you more attractive to the opposite sex, your behavior would change. You would feel more confident and secure, you would feel better about yourself, and you would be more flirtatious and happy—all of which would increase your attractiveness to others and no doubt boost your sex life. None of this has anything to do with magical pheromones; all of it has to do with your self-confidence and happiness.

Before going further into the pheromone question, it is instructive to take a step back and try to understand the origin of our obsession with pheromones. To do this it is useful to know how the word *pheromone* entered our lexicon in the first place. The word *pheromone* was coined in 1959 by Peter Karlson, a German biochemist, and Martin Lüscher, a Swiss entomologist. It is derived from the Greek *pherein*, meaning "to carry," and *hormon,* meaning "to excite"—in other words, "carrier of excitement." Karlson and Lüscher used this word to describe what they were witnessing in their insect lab, that a chemical substance released by one insect—in their case termites—seemed to affect the behavior of other termites around it. They coined the word *pheromone* to describe the general phenomenon that a chemical released by one animal could affect the physiology or behavior of other animals within its species. Simply put, pheromones are chemical communication, and they are highly important for animals, like the social insects, who use chemical signals as their primary mode of communication.

Pheromones also convey important information for many noninsect species, including primates. The reason for the bawdy connotation of pheromone in our modern parlance is because some of the most important messages pheromones carry are communiqués about reproductive status and availability. For example, androstenone, a pig pheromone, turns a sow's attention to mating and nothing else, and it induces her to assume the sexually ready position. This automatic reaction has been exploited by pig farmers wishing to spare the expense of keeping male studs. There is a commercially available spray of androstenone called "Boar-Mate" that, when misted at a sow, eases the process of artificial insemination. A male rhesus monkey will even ignore an amorous female in heat if he cannot detect the pheromones that signal her fertility. The fact that other mammals produce and react to pheromones that play an indispensable role in their sex lives is why people in the fragrance industry hold out hope for a human sex pheromone. If such a chemical aphrodisiac could be discovered and bottled, it would be the biggest thing in the history of cosmetics and fragrance.

Because many pheromones involve chemical secretions that are "smelled," or have a smell, this has led to the erroneous conclusion that pheromones are odors, which they are not. Pheromones are chemicals that *may* or *may not* be smelled at all. It is also the case that pheromones are typically not picked up and processed by the olfactory system, but rather by a separate structure called the *vomeronasal organ (VNO),** which connects to the accessory olfactory bulb, an independent structure from the main olfactory system. The VNO is located above the roof of the mouth and evolved to detect large molecules and molecules that are dissolved in liquid, which is why licking various body parts—as dogs do when they greet each other—is a key way for pheromonal information to be received. We can only "smell," with our nose and main olfactory system, small airborne molecules.

All vertebrates that have been documented to use pheromonal communication use their VNO as the primary system for detecting them. One major problem for creating our billion-dollar pheromone industry is that we do not have a functioning VNO. Human embryos may have a VNO, but after birth this tissue disappears. There continues to be controversy surrounding this issue, but overwhelming evidence points to there being no functioning neural tissue in humans that corresponds to the VNO of other animals. Moreover, although it may be possible for the main olfactory bulb to process pheromones, the accessory olfactory bulb to which VNO nerves normally project has not been found in humans. So what does this mean for us and pheromones?

---

*Also known as Jacobson's organ.

Despite the absence of a functioning VNO and accessory olfactory bulb, there is evidence that humans do demonstrate at least one pheromonal response. This is the often remarked-upon "coincidence" that women who live together over time begin to get their periods at the same time of the month. This coincidence is known as the *McClintock effect* after Martha McClintock's observation that women in the same dormitory began cycling in synchrony over the course of a college semester.* McClintock went on to do a number of experiments that demonstrated that certain chemical excretions from one woman in a group were capable of altering the menstrual cycles of other women with whom she was in contact.† Interestingly, it seems that one woman in particular, known as the "driving female," brings other women in line with her menstrual cycle. This driving female, however, is not predicted on the basis of any physical or social traits. The most popular, prettiest, loudest, sexiest, most affectionate, or smelliest sorority sister is no more likely to be the driving female than the one with the least of these attributes. It is not known what singles out one woman in the group as being the driving female, but ultimately the signaling of her cyclicity must be stronger than that of the women she is affecting. Since

---

*The McClintock effect is a different type of pheromone response than the "signaling pheromone" that causes a sow to immediately assume the posture for intromission. Rather, the McClintock effect falls into the class of pheromones called "primer pheromones," where effects transpire slowly and influence physiology, not behavior. See McClintock, M.K. (1971). Menstrual synchrony and suppression. *Nature, 229,* 244–245.

†Women using birth control pills or other hormonal birth control methods do not experience the McClintock effect.

we have few secretions that could be transmitted easily to one another through group living other than sweat, and since sweat has a smell, there has been an assumption that the scent of sweat is the pheromone for triggering menstrual synchrony.

The McClintock effect is the best evidence for pheromonal communication in humans, but there is still the problem of how it is that women can pick up this pheromone if we are missing the organ to detect it. In McClintock's experiments a sweat solution was applied to the skin above the upper lip—that is, right under the nose—of the women being tested for its effects. Because of McClintock's under-nose manipulation, it has been presumed that smell must somehow be involved. However, the sense of smell is not how other animals detect most pheromones, so it is possible that there is an alternate explanation.

Rather than being smelled or perceived through the olfactory system, I propose that the chemicals responsible for inducing menstrual synchrony are transmitted directly through the skin. That is, the sweat of the donor female is absorbed through the skin of another woman through touch (sweat-to-skin contact), which over time enters her bloodstream, causing changes to her hormonal system that are in synch with the donor woman. Imagine women living together: touching hands, brushing by each other in the hallway, and otherwise being in physical sweat proximity. In this naturalistic setting, direct physical contact may not even be necessary and simply borrowing your roommate's sweater or picking up a book or coffee mug that she recently held may be sufficient. If this is the case, putting a sweat solution on the skin under the nose is irrelevant. I would argue that McClintock could just as well have put it behind women's knees and gotten the same effect. Although my alternate explanation has not been tested, there is evidence that some aromatic chemicals can cause physiological changes through transdermal absorption.[14] My explanation also appeals to parsimony because it requires no "smell" and obviates the problem of our not having a VNO or accessory olfactory system.

In addition to Martha's eponymous effect, there are several "quasi-pheromone"* reactions that have been observed in humans, and they also all appear to be dependent on sweat. The following is a classic example.

Ashley had begun to worry that she might have a gynecological problem. Her periods had become so irregular that she could be weeks late or they arrived with barely a fortnight between them. She knew she wasn't pregnant and that she hadn't contracted any sexually transmitted diseases because she had been single and celibate for almost two years. Then Ashley met Daren and within a few months of dating her periods were as regular as clockwork.

What happened when Ashley met Daren? Our animal cousins shed light on this phenomenon. Exposing female rodents to

---

*There is variability surrounding these sweat/body-odor observations, and the mechanisms and specific chemicals underlying them are currently unclear. Therefore, the signals that elicit these physiological changes are not considered true pheromones yet and instead are referred to as "quasi-pheromones."

the odor of a male will cause the females' estrus cycle* to synchronize. A similar mechanism may be at work in us when a woman has the kind of contact with a man that would enable his scent and sweat to be perceived. George Preti and Chuck Wysocki from the Monell Chemical Senses Center indeed found that secretions from the underarms of men could alter the timing of ovulation and the length of menstrual cycles when applied to the skin under women's noses and could bring irregular menstrual cycles into normal length.[15] Contact with Daren's sweat is why Ashley's periods had become regular.

Julie Mennella from the Monell Center has also found that the sweat from breast-feeding women's underarms changes other women's menstrual-cycle length and increases their sexual desire. Furthermore, the sweat from breast-feeding mothers seems to prompt the production of breast milk in other mothers, especially first-time moms. It even seems that baby sweat/body odor can increase a woman's fertility. You have probably heard stories of women who, after trying every available method to boost their fertility, finally give up and adopt a baby only to suddenly find themselves pregnant. The latest research suggests that baby is actually what does the trick; in other words, baby scent begets babies.

In spite of growing evidence that chemicals in human sweat have physiological consequences for others, particularly women, there is no evidence yet of a viable human sex pheromone. However, there is one glimmer of hope. Androstadienone is a

---

*The estrus cycle in rodents is a fertility indicator that is similar to the human menstrual cycle.

derivative of the pig pheromone androstenone and is also found in human sweat. Recent studies suggest that women feel more positive, particularly around men, when they are exposed to androstadienone, albeit at much higher levels than are in human sweat. But being in a good mood and hauling the nearest man to a motel room are not the same. Our sexual and romantic responses are variable and complex and though biology is a powerful underlying force, there are many social and cultural factors that can usurp and supersede it. In the context of biology, think of how strange it is that many "arranged marriages" actually work out to be durable and loving, child-producing liaisons, and *interpersonal chemistry* is not even remotely considered. The myriad social and emotional complexities that go into human mate choice are ultimately the reason why finding and bottling a surefire human aphrodisiac, pheromone or otherwise, is likely to remain a commercial fantasy.

CHAPTER 6

THE ODOR OF THE OTHER

The lower classes smell.

—GEORGE ORWELL

During my interviews with Jessica Ross, I learned that since suffering from anosmia she had become extremely anxious about how *she* smelled. She was obsessed to the point of paranoia about whether or not her body odor was bad and if her clothes smelled unclean. She admitted: "If anyone ever gives me an odd look, I immediately assume it must be because I smell. I've started showering at least twice a day, and I wash my clothes as soon as I've worn them." It may sound extreme, but Jessica's fears are very common for people who have become anosmic. But no matter how clean we are, none of us can escape "our smell."

Where does *our* smell come from? Human underarm sweat consists of secretions produced by three types of glands. One gland in particular, the apocrine gland, which becomes fully developed during puberty, produces a secretion that is very rich in protein. Jessica Ross, like the rest of us, is shackled by the fact that despite the most scrupulous scrubbing, our underarms are naturally populated by colonies of various types of bacteria. These bacteria feed on the protein on our skin, and as they digest they release gases reflecting their meal. The specific proteins that populate your underarm are genetically determined and reflect your individual MHC gene profile, which is why each of us has a unique smell.

The main compounds that give underarm odor its bouquet are carbon chain acids, but other chemicals are present, too, in particular one that has recently received a lot of press—the pig pheromone androstenone, and its derivative androstadienone, which may be a "quasi-pheromone" in humans. Men typically, but not always, produce higher quantities of androstenone in their apocrine sweat than women do, and differing amounts of androstenone give someone a characteristically "male" or "female" scent—or more accurately, a "stronger" or "weaker" body odor. In blind sniff tests, raters typically categorize body odors that smell weaker as *female* and stronger as *male,* though they are often wrong. Our apocrine glands also emit secretions in response to sexual and emotional arousal and are responsible for what we call the "smell of fear."

Robert stepped out of the hot sun and into the unair-conditioned travel office on the small Greek island he was visiting. It was 105 degrees in the shade and after walking for less than fifteen minutes, Robert's face poured sweat, his hair was a wet tangle, and his T-shirt clung transparently to his drenched chest. The clerk stared at him—"Did you just walk through a sprinkler?" she asked with a slightly nervous laugh.

No, Robert hadn't just walked through a sprinkler; his body was just extremely efficient at keeping him cool. The two other glands that make us sweat are our eccrine and sebaceous glands. The function of our eccrine glands is to regulate our body temperature by eliminating water for evaporative cooling. We are born with our eccrine glands fully functional, and they cover our entire body surface. Eccrine sweat drenches us when we get hot either from the ambient temperature, from physical exertion, or both. One of the problems as we age is that our bodies become less efficient at producing eccrine sweat and consequently at cooling us off. This is one of the main factors contributing to the spike in deaths you read about during heat waves. When you are running beside someone on a treadmill at the gym and the more the person's forehead glistens, the more you smell garlic or some other strong volatiles he or she consumed at lunch, it is because odors from volatile molecules on the skin surface, such as recently consumed food, are in eccrine sweat. Eccrine sweat itself does not smell.

Secretions from our sebaceous glands, which are primarily concentrated on the upper body, forehead, and scalp, do not smell either, but they provide the moisture and the substrate needed for bacteria to grow. Like apocrine secretions, the scent from areas of our body rich in sebaceous glands is due to the bacteria living on the skin surface there. The smell of the top of our heads is largely due to sebaceous gland secretions, and though babies do not have fully functional sebaceous glands, it is commonly believed that there is a typical and distinctive "baby-head" scent. Surprisingly, this has never been scientifically studied and though most people think that Laurie and

Paul's little heads smell the same, mothers can easily distinguish the scent of their own newborn from another by nothing more than a sniff of their crowns.

Our responses to the scents of one another profoundly influence almost all our social interactions and relationships. The feelings that emerge when we catch a whiff of someone else can range from unconditional love to repugnance and prejudice. The primeval *like-dislike* that we get from a scent is not reserved for roses and skunks but extends to everything we smell, especially one another.

SECURITY SCENT

Martha and Dan are almost ready to go out for the night. They haven't had any real alone time in months, and Dan can't remember the last time he took his wife to their favorite restaurant. Martha zips up the sexy black dress she is thrilled she can squeeze back into and Dan pulls out a new leather jacket he has been looking forward to wearing. The doorbell rings and Martha steps into the hallway. The babysitter has arrived! Susan says hello to Martha and Jake, the one-and-a-half-year-old she is going to take care of this evening. Martha gives Susan her cell-phone number and calls Dan to join her at the front door. Everything seems to be going just fine. But as Martha puts on her coat, Jake flings himself to the floor and a meltdown erupts like he has never had before—screaming and wailing and choking with tears that will not stop unless Mommy does not leave. After twenty endless minutes of trying in vain to calm Jake, it is clear that the only hope for tranquility is to cancel their date and stay home. As Susan leaves, Dan gives Martha a look that groans irritation, and she nods in frustrated agreement.

Separation anxiety is very common among young children and is typically at its peak when they are between eighteen and twenty-four months of age. When it is severe, a child's distress at his parents' departure can lead to missing important appointments, dinner dates, or work and can make parents feel angry and guilty. The dream of parents like Martha and Dan is to feel that they can leave Jake with their trusted babysitter and that he will be able to spend a calm and restful night. But Jake and children like him do not feel comforted and relaxed unless Mommy is nearby. What could possibly substitute and create the *feeling* of Mommy without the physical presence of Mommy?

A few years ago I had the idea to develop a garment that could be used to soothe babies and young children who were distraught by their mother's, or primary caregiver's, departure. The garment would be a soft, plush, cotton shirt that the mother would wear for several hours in direct contact with her skin and that then would transform into a blanket to swaddle or otherwise be nestled with her child when she was gone. The garment would be cut and attractively fashioned in such a way that the two permutations were easily and quickly reversible. I called this invention "The Mommy-Scented Convertible Cover."

I spoke to a patent lawyer about this idea, and though she thought it was a bit offbeat, she didn't dismiss it entirely. To our surprise, however, after searching through the patent file records, she discovered that such an invention already existed! In fact it had been developed by a nurse in Minnesota several years before.

Regina Sullivan and Paul Toubas from the University of Oklahoma studied newborns on a maternity ward separated from their mothers for their responses to the body odor of "Mommy." At the time of testing the infants were either (1) calmly awake, (2) crying, or (3) sleeping. They were then presented with either (a) the hospital gown their own mother had just been wearing, (b) the gown that a mother of another newborn in the maternity ward had been wearing, (c) a clean gown no one had been wearing, or (d) nothing. Sullivan and Toubas saw that when crying infants were exposed to the gown that their *own* mother had recently worn they stopped crying. Calmly awake babies also showed extra interest and seemed to be happier when exposed specifically to the gown that smelled of their own mother.[1]

The lightbulb I had for "The Mommy-Scented Convertible Cover" came from knowing about Regina Sullivan's research and the potency of emotional learning with odors. Mommy scent is soothing because the emotions associated with Mama become attached to her scent such that her scent acts as an emotional proxy for Mama herself. A keen observer on a maternity ward would quickly recognize the amazing succor of Mommy's smell. The nurse who figured this out before me must have known firsthand what soothing effects Mommy's scent can have on a distressed infant. I don't know if this nurse has ever tried to commercialize her patent, but I am sure she would be successful because there are plenty of parents like Martha and Dan.

NOSING OUR WAY TO THE NEST

Something besides emotional learning must be taking place for infants to be able to recognize their mothers so quickly after birth, because they can do this in less than three hours after being born, too soon for any meaningful emotional bonds to have developed. So how do infants know who their mommy is, and how do they learn this so fast?

It turns out that there is a basic physical component involved in how and why infants learn the scent of their own mother so quickly, and why infants have more difficulty learning who their daddy is, or anyone else who might have cuddled them right after birth. The reason is because the scent of Mama is very much like the scent of baby herself. The amniotic fluid that has been the baby's home for the past nine months contains chemical constituents of both baby and Mama. The scent of amniotic fluid is more prevalent on a woman's skin when she has recently given birth. Newborns only a few hours old have an affinity for the smell of any amniotic fluid. It is that *familiar* smell they recognize so readily in their mothers. Newborns also need to have direct physical contact with their mothers in order to speedily learn Mommy from Auntie. This is one of the distinctions between bottle-fed and breast-fed babies. Breast-fed babies who have nestled their noses directly against their mother's breast and armpit have intimate contact and knowledge of their mother's signature scent.

Bottle-fed babies who have not had as much nose-to-skin contact take significantly longer to learn Mommy's scent. For similar reasons, babies don't learn the scent of their fathers as readily, unless they have intimate and regular contact with his skin.

THE PARENTAL NOSE

Parents, especially mothers, are also excellent at learning to recognize the scent of their own babies, but unlike their infants, they can do so with only the briefest nose-to-skin contact. In one study of seventeen women who had given birth by cesarean section, 80 percent were able to identify their own two-day-old infants from an unfamiliar infant of the same age, by T-shirt smell alone.[2] This is noteworthy because the women who had given birth by cesarean had very little physical contact with their infants (less than two and a half hours) and yet were still able to recognize their own newborns' distinctive scent.* In fact, it seems that mothers need almost no time at all to learn their infants' unique odor. Scientists at Hebrew University in Jerusalem found that 90 percent of women who had experienced no more than an hour, and some as few as ten minutes, of contact with their newborn were able to identify their own baby

---

*The reason for the strikingly brief contact that moms had with their newborns was because in the mid-1980s C-sections were performed only when acute medical conditions warranted them, and maternal trauma and postrecovery hospital conventions were more severe than they are today. These factors minimized the time a mother could spend with her newborn.

from another on the basis of smell alone.[3] This high rate of recognition was still evident when mothers were blindfolded and the babies had been thoroughly washed. The speed and readiness by which mothers learn to recognize their infants makes good evolutionary sense. It is clearly adaptive for mothers to quickly learn who their infants are since, in case of threat, the mother is a lot more capable of finding and rescuing her baby than a newborn is of finding its mother.

Fathers can also recognize their own infants by smell, but mothers are better at it, and it isn't because mothers spend more time with their children. It's because they are women. Steven Platek and his colleagues at SUNY Albany compared eighteen male and thirty-two female college students for their ability to recognize their own body odor from a set of samples that included four strangers. Over half of the women in the study could correctly identify their own sweat, but only one male student was capable of recognizing his sweat.[4] Since women can recognize their own scent better than men, they can use "self" as a more effective referent for determining who their children are.

Self-similarity is a major factor in parents' ability to recognize the scent of their children. The genes that give us our unique odor signature come from our MHC, and each parent shares approximately 50 percent of these genes with each of his or her children. Children will smell more like their parents than their grandparents or the kids next door. Parents aren't the only ones who can smell this resemblance. Adults who had never met a group of mothers and their children could correctly match up the right mom with the right child on the basis of body odor.[5] Brothers and sisters also share 50 percent of their genes, and they, too, can effectively use smell to recognize kin from kith, even after long periods of separation. In one study, forty siblings who lived in different cities and who had not seen each other for up to thirty months were given T-shirts to smell that had been worn either by their own brother or sister or an unrelated stranger of the same age and sex. Out of the nineteen male and twenty-one female participants, twenty-seven were able to correctly select the T-shirt worn by their own brother or sister. This level of correct identification, 68 percent, is statistically greater than what would be expected by chance and indicates that there is something especially recognizable about the smell of kin.[6]

Mothers and fathers also have no trouble accurately distinguishing between the body odors of two of their children living in the same house and eating the same meals, as long as they aren't identical twins. Even tracking dogs can't distinguish between identical twins on the same diet, because they have the same MHC. Diet is in fact a crude way of recognizing members of your own family. Families who regularly dine on meat and potatoes will smell different from families who are gourmands of curry and garlic. But shared diet and environment can only go so far, because though husbands and wives may eat together and live in the same house, they are not usually genetically related and therefore should not smell much alike. And indeed other people don't think they do. Blind testers asked to match up husband and wife pairs simply by sniffing their T-shirts were unable to make the correct matches.[7]

Even though strangers can't correctly match up husbands and wives on the basis of their scent, spouses, best friends, and other close but unrelated people can easily recognize each other by their body odor. The reason intimates can do this but strangers can't is because of the *learned* familiarity of each other's scent that has taken place over a period of physical closeness and shared experiences. However, when changes such as new diet, illness, medication, fragrance products, or anything else that affects the person's odor profile occur and the partner isn't aware of or hasn't become familiar with the change, recognition ability severely falters.

COMFORT SMELLING

Feeling lonely, Carl picked up the blouse that Judith had worn the day before she left for a month in Paris. Smelling her aroma and feeling the soft, smooth cloth against his face gave him a feeling of happiness he had been badly missing. Discovering an unexpected comfort in this feeling, Carl took Judith's blouse to bed with him that night and slept deeply for the first time in days.

Don McBurney, a psychologist at the University of Pittsburgh, has been interested in what he calls "olfactory comfort" for several years. In one study his research group found that women frequently smelled and slept with the clothing of loved ones they were separated from, and the number one reason they gave for doing this was because they *felt comforted by the smell*. In another study, McBurney and his colleagues found that men also smelled and slept with the clothing of a departed romantic partner nearly as much as women did and for the same comforting reason, but rarely with the clothing of a non-romantic significant other. In contrast, women admitted to sleeping with and smelling the clothing of people they felt close to across a varying set of relationships.

One of the most interesting findings to come out of this study was that the likelihood a woman would seek smell comfort from the clothing of someone she was not romantically involved with was highly correlated with the degree of genetic relatedness between them. Twenty-five percent of the women in the study said that they had deliberately smelled the clothing worn by a first-degree relative, such as a parent, child, or sibling, to make them feel better when they were separated from them. But this number dropped by half when it came to smelling garments worn by grandparents, aunts, and uncles and was almost zero for great-grandparents and cousins.

The finding that genetic relatedness correlated so well with the act of *comfort smelling* may simply be because one is more likely to be emotionally and socially closer to first-degree relatives than extended family members. Social bonds clearly play a role in comfort smelling because McBurney's group found that the likelihood of sleeping with a friend's sweater was about the same as with a grandparent's. But it is also possible that body-odor similarity, due to a greater proportion of shared MHC genes with immediate family members than distant ones, makes the scent of close relatives inherently more familiar and hence soothing to us.

SCENT OF AGE

The scent of a cleanly washed baby is often touted as a "universally good smell."* Since pleasant smells make us feel good, and the odor of freshly washed children is supposed to be nice, is it possible that children's body odor can improve our mood? Although this connection may seem like a leap, Denise Chen at Rice University in Texas investigated this very idea. Chen tested over three hundred male and female college students for their emotional reactions to the body odor from six sources—five-year-old boys and girls, male and female college students (average age, twenty), and elderly men and women (age seventy-one and older)—to see whether any of the scents could produce mood changes among those exposed to it. To her surprise, Chen found that although the body odor of young children was considered most pleasant, it had little effect on the participants' emotions. Rather, the body odor from older women was best at improving mood and actually made students who self-rated as "depressed" feel substantially better. "Older woman" scent, however, was not evaluated as particularly pleasant, but it was rated as highly familiar. Grandma's scent makes us feel good because her scent is comfortingly familiar.[8] This conclusion, however, begs the question: If the grandmothers were unrelated to the students, how could their scent be familiar? Is there a general scent of "old age"?

The data and the dogma suggest that as we get older our

---

*My use of quotes is because a "universally good smell" implies that the response to it is innate, which I do not agree with.

body odor becomes more disagreeable, particularly in old age. Researchers at the fragrance and cosmetics company Shiseido, in Japan, recently conducted a study where twenty-two men and women ranging in age from twenty-six to seventy-five wore a T-shirt to bed for three nights. The T-shirts were then subjected to chemical analysis, and it was found that those belonging to the nine subjects over the age of forty contained a greater amount of a certain chemical with a "characteristically unpleasant greasy and grassy odor."[9] Therefore, the body odor of older people may share common qualities.* But whether this scent is unpleasant is another matter. In the Japanese study, the T-shirts themselves were never rated for how pleasant or unpleasant they smelled. George Preti, America's best body-odor chemist, has not been able to replicate the Japanese findings and claims that allegations of bad body odor among the elderly are anecdotal and to the extent that they are true are due to poor hygiene or the neglect of being institutionalized. The reason why the Japanese researchers were particularly judgmental about the chemical characteristics of older sweat may be because Asians are more negatively predisposed to the qualities of body odor than North Americans are. A strong body odor in Japan is so rare that at one time it was a sufficient disability to disqualify someone from military service.

---

*The chemical components of the body-odor samples in Chen's study were not examined so we do not know whether "grandmother odor" relates at all to the distinctive aromatic compound found in the body odor of the elderly as reported by the Shiseido chemists.

STENCH OF THE STRANGER

During the nineteenth century, scientists were earnestly classifying everything, including human odors, by sex, age, race, and even hair color. Brunettes were said to smell pungent and blondes musky.[10] There are no current data on a hair color connection to scent, but there is reliable evidence that age, sex, and race are involved in some general body-odor qualities.

Asians, for example, have far fewer apocrine glands than Caucasians do, and they also have less body hair. These two facts collude to make the body odor of Asians significantly fainter than that of Europeans. The Japanese anthropologist Adachi quipped in a book published in 1903 that "the yellow race does not smell at all." By contrast, Adachi wrote that Europeans were always noticeable by their body odor, even immediately after bathing, and that the smell of white people was quite foul. Even the renowned Sanskrit sex manual, the *Kama Sutra*, makes reference to the repulsive stench of the hairy white man. It is always a great surprise to white Europeans to learn that their body odor is regarded as inherently offensive by anyone, especially since they have so frequently used body odor to justify their own superiority.

> Among beliefs which profess to show that Negro and white people cannot intimately participate in the same civilization is the perennial one that Negroes have a smell extremely disagreeable to white people . . . white people generally regard this argument as crushing final proof of the impossibility of close association between the races.[11]

This quotation is from a book published in 1937, but the ignorance of this statement is one that has outlived previous-era intolerances. Qualities of human body odor are boldly used today as shameless justifications for racism. Former prime minister of France Jacques Chirac, in seeking to win a share of the "anti-immigrant" vote, openly declared his sympathy with the French worker for "having to put up with the noise and smell of the immigrant family living off welfare next door." The foreignness of *smell* along with the foreignness of the "other" goes deeply toward ingraining ethnic denigration and xenophobia.

Different ethnic groups do not smell alike for many reasons that have nothing to do with race, genes, body hair, or sweat glands. As previously discussed, you are what you eat when it comes to eccrine sweat, and cultures with predilections for certain spices in their cuisine will also wear them on their skin. In addition to secreting various food aromas in our sweat, attitudes and the ability to eliminate sweat scent also vary by culture and have very little to do with personal hygiene or social class. I was once shocked by the *air* on an Air France flight to Paris, as the well-heeled passengers walked by me in their expensive garments and perfumes, bringing with them a potent and distinctly unwashed aroma. When I commented on this later to a body-odor chemist, he explained that it wasn't a fashion statement or lack of bathing, but rather a chemical hindrance in French detergents that prevents the residue of fatty acids from sweat in clothing to be completely removed. Many clothing detergents in Europe do not contain chemicals that effectively dissolve the fatty acids that carry the smell of our sweat. This means that Europeans can smell strongly of body odor even when their clothes are fresh out of the wash and they are fresh out of the shower. Since we adapt to our own odor, this "body odor" would likely go unnoticed by most of those emitting it. North American society is also much more deodorized than European and less charitable toward what are perceived as unclean smells.

BATH AND CLASS

In addition to justifying ethnic bigotry, body odor has been used to rationalize social-class delineations. From the Middle Ages in Europe until fairly recently, bathing was actually widely regarded as a health hazard. So entrenched was the fear that submersion in water led to ill health that during the Black Death, bathing was banned in western Europe. The reason getting wet and soapy was considered such risky business was because it was thought to make the body soft and moist and hence vulnerable to the prevailing unhealthy, "smelly" air, which was believed to be directly related to disease. For most people, cleaning oneself was almost exclusively restricted to washing one's hands, face, and occasionally, clothes. But by the beginning of the nineteenth century this attitude began to change, particularly in the New World. In 1860, Boston, with a population of 177,840, boasted 3,910 portable and mostly unplumbed baths—which was impressive for the time, considering that Albany, the capital of New York, had only nineteen.[12]

In addition to being rare, baths were not cheap, and so at about this same period, personal bathing began to be recognized as something that money could buy, but poverty could not, and hence the foul and the fragrant became a demarcation for rich and poor. Reversing the moral order of earlier days, when St. Francis of Assisi declared dirt to be a sign of holiness, now the more noticeable the odor of any kind, the greater the moral promiscuity was assumed.

At the turn of the twentieth century, the poor not only "smelled unclean" because they labored and could not afford private baths, their cramped living conditions also meant that the odors of their kitchen, bedroom, and toilet tended to mingle indiscriminately in their crowded houses. One's scent, therefore, made one's social class obvious and was used as justification for some of the worst classist attitudes. This was explicitly and succinctly put by George Orwell, who in the late 1930s wrote: "The real secret of class distinction in the West can be summed up in four frightful words . . . the lower classes smell."[13] Orwell went on to explain that this scent distinction was a rational and insurmountable cause for class discrimination, though this "smell" was clearly colored by context and connotation. Orwell admitted that his repugnance for "lower-class odor" was not necessarily due to personal hygiene, because even the cleanest and most fiercely scrubbed servant still *smelled*.[14]

Early in the twentieth century, in addition to disdain for body odor, breath odor was added to the mix of possible causes for social ostracization, and commercial opportunities for deodorizing mouths were seized upon. Listerine, which hit the market in the 1870s, was originally sold as a general antiseptic for homes and hospitals. But in 1920 the Lambert Pharmaceutical Company, which made Listerine, saw a new opportunity and reinvented Listerine as a mouthwash. In fact, Lambert Pharmaceuticals coined the term "halitosis" in 1921 as a medical definition of bad breath. At this turning point, Listerine forgot its hospital base and began aiming its ads to frighten the young and single, with the goal of instilling paranoia that without fresh-smelling breath one was in danger of dire social consequences. Here is an example ad line from that campaign: "She was often a bridesmaid but never a bride. And the secret her mirror held back concerned a thing she least suspected—a thing people simply will not tell you to your face. . . ."

The tactic worked and the annual profits for the Lambert Pharmaceutical Company went from $100,000 in 1920 to $4,000,000 in 1927. Nothing had changed about the Listerine formula except for its image.[15]

LAW AND ODOR

You enter a crowded elevator and are instantly assaulted by someone's exceedingly pungent body odor. *Who hasn't showered in a month? How rude that they don't care how offensive their body odor is!* You scan around angrily, *Why does this person, whoever they are, smell so bad?*

A FEW YEARS AGO I received a call from a lawyer wondering what I knew about extremely bad body odor. The lawyer had been retained by an ad agency where one of the employees' body odor was so unpleasant that his coworkers were refusing to come to work, and no clients would deal with him. Apparently his body odor created such a stench that to be in the same room with him was nearly intolerable. The employee had been spoken to about this problem but claimed he could do nothing about. I was also once contacted about a case in which a landlord wanted to evict a tenant because his body odor was so unbearable that his neighboring tenants began moving out and demanding damages from the landlord. In both cases, the claimants wanted to know if I could supply scientific reasons or precedents to fire, fine, or evict someone because of how they smelled.

There is a disease called *trimethylaminuria,* or "fish-odor syndrome," where an individual is unable to break down the compound trimethylamine and as a consequence has a very pronounced fishy smell in his or her breath and body odor. This syndrome is genetic and has been pinpointed to a recessive mutation. Unfortunately for those with the illness, there is no known cure at present. But this disease is rare, and since it is present from birth, it is unlikely to be the cause of most coworker or cotenant complaints. Being obese can also produce metabolic abnormalities that result in abnormal and intense body odor and these symptoms will be exacerbated by poor diet and hygiene. The "smelly" people whose bosses and landlords were trying to get rid of them may have been obese, had poor hygiene and housekeeping habits, or possibly suffered from trimethylaminuria, or some combination thereof. I cannot say for sure because I opted not to become involved in these cases.

SCENT AND THE CITY

From the very first cities to the end of the Industrial Revolution, odor pollution was at a magnitude we can barely imagine. Even with fumes from cars and factories, most urbanites in the first world today live in a relatively scent-free atmosphere. During the Renaissance, city streets in Europe were used as conduits for waste of all kind: food remains, human and animal excrement, blood and entrails from slaughtered animals, dead cats and dogs—even human blood from surgeons was casually thrown into the street. The streets themselves were made of dirt, which culminated in liquidy effluents producing a truly stinking sludge. The stench of the city was so bad that smells themselves were often claimed to be a cause of death.

The reason for the "smell" reform of Western cities in the mid-nineteenth century is due to the connection that was realized between filth, germs, and disease and the consequent institutionalization of public sanitation works that developed to get rid of waste and, as a corollary, stench.

The oldest-method known to eliminate undesirable aromas is simply to mask bad odors with a "better, stronger odor," which is apparently why cavemen dragged pine and fir boughs into their caves. Masking odors is also a primary tactic used today, and one famous city has been working hard to improve its ambient aroma in this way for nearly a century. Starting with perfumes in the 1920s, numerous attempts have been undertaken to improve the odor quality of the Paris metro and to overwhelm, with something better, that mélange of garlic,

Gauloises, stale perfume, body odor, and axle grease that wafts through the underground rail system. In the years between 1990 and 2000, various fragrances—one ironically named Madeleine, which had a woody musk base, a touch of vanilla, and a blend of citrus, lavender, jasmine, rose, and lily fragrances, was used as a mask for "eau de metro." The maintenance service mixed the fragrance with detergent and applied it to floors and other surfaces whenever the cleaning machines were used. An advertising campaign went along with the metro perfume to draw attention to the subway service's aroma intervention with posters that asked "Do you smell the difference?"

Apparently, some Parisian riders appreciated the subway's efforts to enhance the air quality, but most thought the added aromas made little or no improvement. Finally, after a five-year research project and almost $3 million, titanium dioxide was chosen as an effective deodorizing method. Titanium dioxide is better known as a compound in sunscreen. The way it works as a scent deodorizer is by forming free radicals that destroy smelly organic compounds when it is irradiated by ultraviolet light.

In the United States at least $50 million is spent annually to control odors in sewers. The simplest method to neutralize odors is with charcoal, which absorbs odor molecules. Other methods include mixing "good" odors with "bad" ones to create a new scent. Indole and skatole are both fecal-sewage scents, but mix them together with a few other chemicals and you get a fine perfume. In fact, some sewage plants use perfume to neutralize the indole and skatole emanating from their smoke-stacks. Other current methods for odor cleansing include burning odors and "chemical scrubbing." At 1,200–1,600°F you can burn complex chemicals into odorless air in a few seconds. Chemical scrubbing breaks down fetid odors into simpler non-offensive or unsmellable molecules.

Another very pungent industry is swine production. Swine production revenues in the United States average more than $12 billion per year with at least 100 million pigs bred annually. This many pigs means a lot of manure, and it is estimated that at least 85 million tons in dry weight solid waste is produced each year requiring treatment and disposal. Odor management has become a crucial issue for the swine industry since complaints are high and dealing with them is costly. Though there do not seem to be any deleterious physical consequences from exposure to the smell of pig manure, nearby residents complain of both health and mood problems, and attempts to close down swine production facilities because of the prevailing scent are continuously being fought.

The characteristic odor of pig manure is primarily due to volatile fatty acids from bacterial breakdown of organic compounds. But it turns out that pig manure can also be *de-scented* by bacterial action. A new technique involving an ion of iron (FeIII) and a particular species of bacteria is currently being explored to neutralize pig manure odor cheaply and efficiently. In a study testing the new method, after five weeks of incubating the pig manure with the bacteria and iron mixture, no significant unpleasant odors could be detected.[16] This means that biological techniques can be used to effectively reduce or eliminate noxious scents. The key to odor elimination is knowing exactly what chemical you are trying to get rid of. For example, vanilla is able to neutralize the scent of chlorine, but chocolate is not.

It isn't only people who live by pig farms or sewage plants who need aroma protection. Disaster recovery workers are often faced with overwhelming odors that can stymie their efforts and produce debilitating aftereffects, in the form of odor-evoked post-traumatic stress disorder. When cleaning up the aftermath of Katrina, many rescue workers avoided this risk by smearing a new gel under their noses called "OdorScreen." Unlike other topical products that coroners and crime investigators use that simply overpower all odors in the immediate environment, Odor-Screen, developed by the Israeli company Patus Inc., interferes with the sense of smell itself. OdorScreen's technology is based on the phenomenon of cross-adaptation, where due to the interaction of specific chemicals that share similar sensory channels, olfactory receptor activity can be changed so that the perception of one odor becomes different or blocked. OdorScreen has been developed to be specifically active against a particular set of nauseating compounds and does not interfere with the perception of most pleasant scents. However, since OdorScreen only works against a predetermined set of odors, it is not a guaranteed barrier to all noxious chemical brews.

A BRIEF HISTORY OF PERFUME

The personal wearing of scent was first recorded by the Egyptians, who put flowers, herbs, and spices into wax cones that they wore on their heads; as the wax melted, the aromatic mixture flowed out and perfumed them. The Etruscans revered perfume as well, to the point that Etruscan women were *never* without it. The Etruscan spirit of adornment, called Lassa, is a naked winged female carrying a perfume bottle. She is depicted on the engraved brass mirrors that Etruscan women were buried with to accompany them to the afterlife. The Romans were also great connoisseurs of perfumes, and gladiators are said to have applied a different scented lotion to each area of their body before a contest. But as Christianity rose with its severe and simple attitudes toward adornment, perfume evaporated in the mist. Fortunately, this austere attitude toward self-scenting was not to be a permanent obstacle to the fragrance-seeking nose.

During the Middle Ages, as European cities were becoming more and more crowded and, therefore, more and more fetid, the prevalence of donning perfume was simultaneously becoming ever greater, and by the Renaissance, perfume was truly *reborn.* By the sixteenth and seventeenth centuries, the fetish of perfuming everything was so extensive that even pets and jewelry were daubed with their owner's favorite scents. By the eighteenth century, perfume was enjoying the status of high fashion and the higher one's importance, the better one's fragrance. In 1709, a French perfumer proposed that the different classes should each have a special scent. He concocted a royal perfume for the aristocracy and a bourgeois perfume for the middle classes, but he said the poor were only worthy of disinfectant. The court of Louis XV, king of France from 1715 to 1774, was known as "La Cour Parfumée"* and the aristocracy

---

*Translation "The Perfumed Court."

was expected to wear a different perfume for every day of the week. The French royals were so in love with their fragrances that some may have even died for them.

On June 20, 1791, Louis XVI, Marie Antoinette, and the family retinue were attempting to escape to eastern France where loyal troops were waiting. However, their flight was cut off at Varennes when they were recognized and arrested. Marie Antoinette's addiction to her favorite perfume Le Sillage de la Reine (The Queen's Wake) is suspected by some to have been the fatal mistake leading to their capture. One account is that a vigilant chambermaid observing the queen's traveling chest packed with several months' supply of perfume realized that the royal family was making a break for it and blew the whistle. Another explanation is simply that the queen's distinctively sweet aroma, in sharp contrast to the odors of her squalid fellow travelers, gave her away.

Italian perfumers living in Koln, Germany, created *eau de cologne* in the 1700s. Eau de cologne was originally made from rosemary and citrus essences dissolved in wine and was even believed to be a preventative for the plague. The scent itself became wildly popular and was such a favorite of Napoleon's that he supposedly doused himself with a vial of it every morning.

During Napoleon's reign, fragrances were not gender specific, but by the late nineteenth century, perfume wearing had become gender stereotyped. Sweet floral blends were deemed exclusively feminine, while sharper, woodsy, pine, and cedar notes were characterized as masculine. However, the wave of fragrance fashion was again about to change.

The dawn of the technological age brought with it a conservative attitude toward self-scenting. In the early to mid-twentieth century, men with any credible social position had stopped wearing fragrance and were expected to smell only of clean skin and tobacco. Women of social standing were expected to smell faintly of floral notes, in accord with their muted position in worldly affairs. Only prostitutes and the déclassé could get away with sporting the once-prestigious heavy and exotic scents of yesteryear.

A break in American perfume repression came surprisingly during the otherwise repressive period of the 1950s. Chanel No. 5, created in 1921, was the fifth fragrance in a line developed by Ernest Beaux for Gabrielle "Coco" Chanel. Chanel No. 5 is composed of a variety of floral and woody notes as well as vanilla. It enjoyed popularity in France and Europe after its inception, but became a superstar when it was launched in the United States in the early 1950s and superstar Marilyn Monroe, asked what she wore to bed, replied, "Two drops of Chanel No. 5." Since the mid-1950s Chanel No. 5 has been the most famous perfume in the world, and it continues to outsell almost all its modern rivals today. In 1985, Andy Warhol immortalized the Chanel No. 5 bottle with a series of nine silkscreen prints.

By the end of the twentieth century, varied and potent fragrances were once again fashionable for women, and fragrance for men was again a prestigious fashion statement. Today the metrosexual is the fastest-growing consumer market for fine fragrance. In the United States, men's prestige fragrance sales topped $900 million in 2006, with Acqua di Gio by Giorgio Armani being the current number one seller. Revenue from "prestige" fragrances for both men and women brings in at least $3 billion in sales annually.

Chanel No. 5 is in the scent category perfumers call "floral aldehyde." There are many other current bestsellers that are composed of classic floral blends; however, Chanel No. 5 was also the first fragrance to be created with synthetic, that is, artificial, chemicals. Before synthetics were used in perfumes, scents faded quickly and fragrance had to be continuously reapplied. This produced substantial demand and hence strain on supply. Jasmine is the most popular flower used in fine fragrance, and it takes upward of fifty pounds of jasmine flowers to create just one pound of fragrance oil. The continued demand for natural jasmine is why visitors to Grasse, perfumery's home central (near the French Riviera), will see fields upon fields overflowing with this white blossom.

Today there are also many unusual options for the perfume esthete. Among the more atypical are those available from Demeter Fragrances, which boasts over 150 scents ranging from the playful to the shocking, including: Holy Water, Dust, Playdoh, Funeral Home, Paperback, and Gin and Tonic—the art savvy can even find "This Is Not a Pipe."* Demeter has not restricted itself to the entirely iconoclastic demographic and wisely has capitalized on one of the top trends in fine fragrance today—food—with Eau de Birthday Cake and Sushi. The *Financial Times* recently recognized "food" as one of the top three trends in fragrance development, with "oriental" and "marine" as its competitors. Demeter isn't the only company vying for the nose of the daring gourmand. The Stilton Cheese Makers Association, located in Surrey, England, recently commissioned

---

*Allusion to the surrealist artist René Magritte.

Eau de Stilton as part of its 2006 campaign to encourage people to eat Stilton as part of everyday meals. According to Nigel
White, a company spokesman, "Blue Stilton cheese has a very
distinctive mellow aroma and our perfumier was able to capture the key essence of that scent and re-create it in what is an
unusual but highly wearable perfume that we are very proud to
put our name to."

   If the idea of cheesy perfume doesn't appeal to you, there are
certainly easy and familiar classics within the food fragrance category you can try. Walk into any Body Shop and you will immediately become aware of how fruit scents are pervading our fragrance
predilections; satsuma, strawberry, and vanilla perfume oils are
among the top sellers. Indeed, vanilla is the most popular fragrance ingredient in both the casual and prestige categories.

   The marriage of perfume and the edible is not only in what
we daub on our skin, but now also in what we splash into our
food. Some daring chefs are currently experimenting with and
proposing recipes for using fragrances like vanilla, jasmine, and
lavender to infuse what you serve at your next dinner party.[17]

CHAPTER 7

CRAVING

There is no love sincerer than the love of food.

—GEORGE BERNARD SHAW

Meet Mark and Julie.

Mark is in his kitchen, checking his watch for the third time in fewer minutes. 11:22 P.M.—He knows The Grille is closed but he cannot get his mind off their chicken wings and his unbearably intense desire for them. Fried to a crispy golden brown, slick and dripping with Frank's hot sauce and butter, he imagines sinking one into the chunky homemade blue cheese dip The Grille is famous for. He isn't sure he can make it until The Grille opens again for lunch the next day. Restless, he circumnavigates the kitchen another time. Mark is hungry but not starving. This is a hunt to satisfy a very specific urge. He searches his cupboards and refrigerator once more without satisfaction. Finally he settles on a microwavable cheese and pepperoni pizza. He pops it in—hoping that this will at least take the edge off his craving for those crispy, tangy morsels. Two minutes later he takes a bite. Hmmm. It's OK but—no—it's definitely not doing the trick. He puts the pizza down, frowns, and mentally begins searching for places that might still be open where he can get some chicken wing satisfaction. He would drive at least half an hour to get them—if only he could think of where.

Julie fidgets at her desk, unable to concentrate on the portfolio in front of her. She looks at the carrot and celery sticks in the Baggie beside her computer. No—that is absolutely not what she wants. Her mind drifts. Chocolate. Rich, dark chocolate. More to the point, the dense, flourless, chocolate cake from the bakery down the street. She imagines her fork slicing into one of its generous slabs. The thick and creamy icing giving way to the moist solid wedge below. Her mouth waters. "That's it; I'm going now," she mutters aloud—"I've had enough of this diet. I've got to have that cake!" She grabs her coat and heads for the door. Her heart races in anticipation of the luscious wedge of sweet, moist darkness that she will soon be feeling with her tongue.

Craving is "an intense desire to eat a particular food," says Marcia Pelchat, the Monell Chemical Senses Center resident gourmet and a world-leading expert on food and flavor cravings. The key words are *intense desire* and *particular food*. Food craving isn't just *wanting* to eat; it is intense to the point that you will go out of your way—drive half an hour at midnight—to find satisfaction for the particular food you want. Craving is also not about feeling starved when you just need to eat and any food will do. When we crave a food, it is for a very specific item—chicken wings, not pizza; flourless chocolate cake from the bakery down the street, not a Milky Way bar. The reason we can't be satisfied with any old variation on the theme of our desired food is because our memories are involved in craving. A sensory template that knows whether it has been satisfied or not. Pizza doesn't match Mark's sensory template of chicken wings, which is why it doesn't quell his urge.

In reading Mark's story you might think you could substitute crack cocaine for chicken wings and get a similar description. You would be right. Food craving and drug addiction are very much alike. The neural structures that are involved in cravings for drugs and alcohol are in and around the limbic system, our brain's emotion and motivation control center. Neuroimaging studies have shown that when drug and alcohol addicts crave their substance of abuse, the specific areas that light up include the amygdala, hippocampus, insula, caudate, and orbitofrontal cortex.

To test whether the same neural substrates were also involved in food craving, Marcia Pelchat and her colleagues put people on a monotonous diet—which causes food cravings.[1] After a few days of consuming nothing but a nutritionally complete but bland diet drink, the participants were brought in for neuroimaging, using fMRI.* The experiment was tailored for each person, and

---

*fMRI is the acronym for "functional magnetic resonance imaging," a variation on the standard MRI medical procedure. In fMRI, brain activity is monitored while it is engaged in a "function" over time.

the researchers had previously discovered the names of two foods that each person "really liked." While the participants' brains were being scanned, the names of these favorite foods appeared on a screen, and the individuals were told to think about the food when they saw its name. The imaging data showed that when the participants were fantasizing about their "really liked" foods, the same areas of the brain that spark when drug and alcohol addicts crave their forbidden pleasures also glowed bright here—in particular the hippocampus, insula, and caudate.

In other words, the reason why Mark is so desperate for chicken wings is because his brain is responding similarly to how it would if he needed to smoke a cigarette or have a hit of cocaine. Eating sweets is inherently pleasurable. Newborns only hours old will smile when sugar is put on their tongues. Abstaining addicts, particularly alcoholics, report much higher than normal cravings for carbohydrates, especially sweets, when they are without their drug. The connection between an overactive sweet tooth and compulsions for other addictive substances is so ingrained and acknowledged that addiction treatment programs recommend using sugary snacks as a means of reducing cravings for drug and alcohol abusers.

Food cravings may be the primal source for all craving, including lust. Julie's story sounds like an erotic scene is about to ensue, were it not for the words *chocolate cake*. The brain region of the insula, one of the three areas activated in the brains of Pelchat's food-craving research participants, is also predominantly active during sexual arousal. Julie's desire for flourless chocolate cake is truly libidinous. The joy of food is the basis for all our pleasures.

Close to 100 percent of young women and 70 percent of young men report having experienced one or more food cravings during the past year, but men and women don't crave the same foods. There is about a 60–40 split when it comes to what kind of foods are craved between the sexes, with women erring in favor of bonbons, and men preferring steak over cake. That is, 60 percent of the foods women crave are sweet—chocolate is high on the list—and 60 percent of the foods men crave are savory, like chicken wings. Not surprisingly, dieters crave foods more than nondieters. In fact, in one of Marcia Pelchat's studies, *100 percent* of young men and women (eighteen- to thirty-five-year-olds) who were dieting reported food cravings.

Food cravings in general diminish for people sixty-five and older, even among dieters. Men and women over sixty-five are also more similar in what foods they crave. While younger women crave sweets more than young males do, among the elderly, women's cravings for sweet treats drops to match craving for sweets among men. Marcia Pelchat has speculated that female hormones may be driving women's craving for sweets, which also explains why during the premenstrual and menstrual phases of a woman's cycle, urges for chocolate and other sweets are reported as especially high.*

It is also possible that the reason elderly adults have fewer cravings in general is because their olfactory abilities have declined. Between the ages of sixty-five and eighty, about one in four people have lost their sense of smell, and after age eighty,

---

*The chocolate–menstrual cycle correlation has not yet been sufficiently scientifically tracked to be conclusive.

half the population has substantial smell loss. Because the loss of olfaction is gradual with age, and the public and medical community pays little attention to our sense of smell, many people are reluctant to report waning olfactory abilities to their family or physicians. Consequently many elderly people have dismissed and/or are unaware that they have such a significant loss. In fact, in studies where the elderly are tested for their olfactory sensitivity, using a device called an olfactometer, where odors are presented without any visual cues, it is typical for the participant to say, "Why haven't you given me anything to smell?" when they have in fact been given many scents to assess that a younger person would have no trouble detecting. The loss of olfactory acuity with age is also why the elderly salt their food so heavily. The presumption is that by adding salt the food flavor will be enhanced, but the only thing that is really intensified is salty taste.

THE TRUTH ABOUT TASTE

What we colloquially call "taste"—the oral sensations from eating and drinking—is really a complex mix of factors that includes taste, as well as the mouthfeel from fattiness and other textures, temperature, and most important, smells. When you bite into a chunk of Godiva chocolate, what you are experiencing is a soft, fatty feeling on your tongue, the mix of sweet and bitter tastes, and most of all the aroma of chocolate.

Although you are probably familiar with four basic taste sensations, there are actually five. They are salt, sour, sweet, bitter, and the most recently discovered one, umami. *Umami* was first discovered by Japanese researchers and roughly translates from Japanese to English as "deliciousness" or "savory." A basic taste means that the perception of that sensation is produced by specific biochemical-receptor interactions. There are five different types of biochemical and cellular interactions, respectively, that produce our experience of salt, sour, sweet, bitter, and umami.

Umami is the taste from the amino acid MSG (monosodium glutamate) and can also be described as the taste of pure protein. In Western cuisine it is best exemplified by meaty broths. Sour taste comes from acidic compounds—compounds with a pH below 7, such as lemon, which is citric acid, or vitamin C, which is ascorbic acid. Salt taste comes from substances that are acid-base mixtures. For sodium chloride, or table salt, sodium is the base and chloride is the acid. Sodium chloride is just one example of a salt, but it is the one we like the best. Sweet taste *generally* comes from carbohydrates. For example, any type of sugar—glucose, fructose, sucrose—is a simple carbohydrate, and they all taste sweet to us. But sweet tastes can also come from sources without any carbohydrate or caloric content, such as amino acids, as in the case of aspartame (NutraSweet). Bitter taste comes from alkaloids—compounds with a pH above 7, such as quinine, the key ingredient in tonic water. The most bitter substance in the world is denatonium benzoate, and it is often added to pesticides and household cleaners to prevent accidental poisoning.

Tastes also interact with one another when they are combined. Salt in particular interacts very positively with other taste sensations, which is why it is such a useful cooking aid.

Salt blocks the taste of bitter and will make something sweet taste even sweeter. You can try this by sprinkling some salt on your morning grapefruit. The salt is actually more effective than sugar because it both blocks the bitter and enhances the natural sweetness of the fruit, though it will also make your grapefruit taste salty.

Certain chemicals found in common household products can also block the receptors for specific tastes. Have you noticed that after you brush your teeth and eat an orange that the orange tastes horribly bitter? This is because there are compounds in toothpaste that suppress the sweet receptors on your tongue so all you can taste from the orange are the bitter compounds. This effect is temporary and can be eliminated by taking a bite of toast before your first sip of orange juice.

Saliva is the medium in which tastes are dissolved and through which they reach the taste-receptor cells. Saliva actually has a slight salty taste due to the presence of sodium ions, though you can never taste it yourself because you have adapted to it. For something to taste salty it has to have more sodium than the saliva in your mouth. The receptors for taste are in cells contained within taste buds. Most people assume that taste is restricted to the tongue but you actually taste with your whole mouth. Taste buds are located in small pits and grooves called *papillae* that are on your tongue, the roof of your mouth, your throat, and the insides of your cheeks. What you see on your tongue are the papillae, not the taste buds. Each papilla has a number of taste buds in it—the average number is about six—and within each taste bud there are forty to sixty taste cells, arranged in segments like an orange. The tongue itself contains about five thousand taste buds. All the taste regions in your mouth combined contain approximately ten thousand taste buds. Everything else you think of as "taste" is due to smell.

One of the major features that differentiates our sense of taste from our sense of smell is that our responses to taste are primarily hardwired and innate, rather than learned as they are with scents. Placing a drop of sugar on a newborn's tongue elicits a smile, while placing a drop of quinine on her tongue will elicit the characteristic "yuck" face, which coincidentally is the same facial expression we make for the emotion of disgust.* A drop of vinegar will make an infant purse its lips as we would when sucking on a lemon. Responses to salt are acquired somewhat after birth and are concentration dependent, with low concentrations inducing smiles and high concentrations less happy expressions. The newborn facial expression to umami has not yet been documented, but extrapolating from the positive attributes umami is given by adults, one can presume that a newborn's face would indicate pleasure.

The basic tastes give us the fundamental codes from which the experiences of eating and drinking arise, and the pursuit or avoidance of them has led to momentous physical, political, scientific, and economic effects. Here are some of the behaviors and consequences that the sensations of bitter, sweet, and salt have taken us to.

---

*This suggests that the origin of emotional disgust has its basis in our physical revulsion for bitter tastes.

BITTER BEHAVIOR

Bitter things tend to be poisonous, but some bitter compounds found in vegetables are extremely healthful, particularly in reducing the risk of cancer. Because there is some survival advantage to consuming bitter-tasting things, it would be most adaptive for us to be wary but not entirely repelled by bitter substances. Nevertheless, a large number of people simply cannot bring themselves to eat their greens because they are much too unpleasant for them. I am one of those people, and we have been dubbed "supertasters" by Linda Bartoshuk, a pioneer and world-renowned expert on taste and oral sensations. "Supertasters" have a gene on chromosome 7 with both alleles in their dominant form. "Tasters" who are sensitive to bitter but not excessively so have one dominant and one recessive allele for this gene, and "nontasters" have both alleles in their recessive form. The population breaks down more or less in thirds for their taster-status genetics.

A standardized test for taster status using a substance known as propylthiouracil, or PROP, was developed by Linda Bartoshuk, and I subjected myself to this test once, which is how I know that I am a supertaster. My mother, however, could have told me this long ago after years of putting up with my refusal to eat the "delicious" endive, watercress, kale—the list goes on. You likely know where you fall in the tasting spectrum yourself by how you respond to these vegetables and others, including escarole and celery, and whether "Campari and soda" is on your drink list. Supertasters have more papillae and hence more taste buds and taste cells than nontasters, and therefore experience all tastes as more intense. We also experience three to four times the burn intensity from eating hot peppers and two to three times more creaminess and fattiness in foods than non-tasters. As a result of our greater taste sensitivity, and potential for quicker satiation on sweet and fatty foods, supertasters tend to have a somewhat lower body mass index than nontasters.

More supertasters are female than male, and more Asians than any other race are supertasters. Chefs are also more likely to be supertasters than not. Perhaps one of the factors that leads a person to become a chef is that their taste world is more intense. In addition to choice of profession, some other behaviors are mediated by your taster status and they are not all benign. As mentioned before, supertasters don't like many leafy green vegetables and therefore don't eat them. This is because leafy greens contain alkaloids, and alkaloids taste bitter. But reduced vegetable intake is a risk factor for cancer. Valerie Duffy at the University of Connecticut, and colleagues at a Department of Veterans Affairs hospital, found that in older men undergoing routine colonoscopies, those who were supertasters had the most colon polyps, a precursor to colon cancer.[2] Similar correlations have been observed for women and gynecological cancers.[3] Therefore, supertaster physiology can induce behaviors that are potentially dangerous (knowing this I try to force myself to eat kale—sometimes—and when I do, I use a lot of salt, because salt makes everything taste less bitter).

Another behavior correlated with taster status is alcoholism. More alcoholics are nontasters than tasters or supertasters. Alcohol has a bitter taste. In order to drink alcohol you have to overcome disliking the taste of bitter, and if you aren't very sensitive to bitter, then you have less to overcome and can consume more alcohol more easily. The amount of alcohol you consume and the length of time you have been consuming it are major predisposing factors for developing alcoholism.

SWEET TRICK

When Christopher Columbus introduced cane to the Old World, sugar was an exotic luxury. Most Europeans had never eaten sugar, but they quickly developed a fondness for it. By 1700, the Americas had become a giant sugar mill and were supplying the British with four pounds of sugar per person per year. By 1800, each resident of the United Kingdom was eating eighteen pounds per year, and by 1900, they were devouring ninety pounds per person per year. This number has steadily risen over the past century. But nowhere has the rise of sugar been more dramatic than in the United States. In 2005, the average American consumed 140 pounds of sugar in various forms, high fructose corn syrup being a particularly nefarious type. This yearly sugar dose is 50 percent more than the average German and nine times more than the average Chinese. Globally in 2005, almost 300 billion pounds of sugar were consumed.

We innately like sweet because sweet signals carbohydrates, which are a survival necessity in a food-scarce world. Yet in spite of the fact that many animals thrive on carbohydrates, no other species is as driven toward the taste of sweet as we are. Cats and chickens can't even taste sugar.

Our lust for sugar coincides with our ever-expanding girth, and in the effort to stem the flow of obesity, artificial sweeteners have been turned to as the next best thing. One problem with artificial sweeteners is that they don't taste like sugar. Another problem is that they often increase rather than quell our appetites. In 1986, an epidemiological study found that women who consumed artificial sweeteners gained weight. That same year John Blundell, an English expert on weight regulation, published a provocative article in the medical journal *Lancet*, stating that aspartame* actually increases appetite. He and his colleague Andrew Hill observed that after consuming products containing aspartame, individuals increased the amount they ate in a subsequent meal. In 2005, the average American ate approximately 24 pounds of artificial sweeteners, nearly double what she or he did in 1980. One would think that this massive consumption of sweet alternatives would mean eating less sugar, but sugar consumption rose nearly 25 percent during the same period. Whether artificial sweeteners are actually making us fatter is a controversial and unresolved topic. This has not stopped the fact that trying to make a truly sugarlike artificial sweetener is one of the holy grails of the taste industry.

The biggest problem manufacturers admit to having with artificial sweeteners is that they are just not as good as the real thing. In 1981, the NutraSweet Corporation launched aspartame, which is the combination of two common amino acids, phenylalanine and aspartic acid, plus a methyl group. Aspartame is ounce for ounce two hundred times sweeter than sugar.

---

*The artificial sweetener sold as NutraSweet.

But aspartame can't be used in cooking or baking because it breaks down when heated, and it also disintegrates over time, losing its sweetness, so diet colas drunk after their "sell-by date" don't taste very good. Many would-be diet drink users have also never been satisfied with aspartame's "sweet" taste.

Since the early 1980s the NutraSweet Corporation has been on the hunt for a better sweetener, and with the help of two French scientists, Claude Nofre and Jean-Marie Tinti at Claude Bernard University in Lyon, has developed the latest contender, "neotame." Neotame is a version of aspartame with a chain of carbon and hydrogen atoms attached. Neotame doesn't lose its flavor or break down when you cook it, and it is about eight thousand times sweeter than sugar. The FDA approved neotame in 2002, and it made its first appearance in U.S. stores in 2006 in products such as Ice Breakers candy and SunnyD Reduced Sugar orange drink. The problem with neotame from a taste perspective it that it is slow; that is, it takes a relatively long time to register on the tongue, and it also lingers a long time after you've consumed a product containing it. Time will tell how well it does on consumers' palates and on the market.

In spite of imperfections, Craig Petray, the chief executive officer of NutraSweet Corporation, claims that the company is no longer trying to create an improved neotame or discover a new calorie-free, sweet magic bullet. Rather, they are focusing their energies on creating blends of preexisting artificial sweeteners that minimize their individual shortfalls and then mixing them with sugar to get as authentic a sugar replacement as possible with fewer calories than pure sugar.

NutraSweet is not alone in realizing that sugar is the key to sweet and that the best artificial sweeteners are likely to be sugar plus "something." Charles Zucker, a molecular biologist at the University of California at San Diego, is cofounder of the biotech company Senomyx. Senomyx is devoted to discovering and patenting receptors for tastes, and it is actively working on the sugar-substitute problem. Zucker and his colleagues have found that each cell in each taste bud is geared to recognize one specific basic taste. Certain cells detect sweet only, others bitter only, and so on. Using a technique called "high-throughput screening,"* Senomyx is identifying specific chemicals that turn various sweet receptors on. When the best chemical "turn-ons" for our sweet receptors are discovered, they will be marketed not as sweeteners themselves but as "sweet taste po-tentiators": chemicals that boost the gain on our perception of sweetness, so that a little bit of sugar plus potentiator will pro-duce the same perception of sweetness as a big dose of sugar naturally would. One of Senomyx's sweet potentiators is known as Substance 951, and only a few parts per million are needed in a can of soda to be able to take out 40 percent of the sugar and have it taste just as sweet. Taste potentiators like Substance 951 can be used in such minute quantities that they won't even need to be listed on labels, so you are unlikely to know when your mouth has been manipulated by them.

---

*High-throughput screening is a general technique currently used in bio-chemical industry and research, where a very large collection of compounds (e.g., hundreds of thousands per day) can be tested against a particular bio-logical system.

BLOOD, SWEAT, AND TEARS

Salt—sodium chloride, the white granular mineral you sprinkle on your steak, corn, kale, and grapefruit—reduces the bitter quality of food, makes the sweet qualities brighter, and adds what food scientists call *roundness*. That is, it takes out the rough edges and makes food taste better.

Salt is part of us and we need it to survive. Our blood, sweat, and tears taste salty because salt seasons all of our bodily fluids. If we didn't consume salt, we would die. This is because we cannot produce sodium or chloride inside our bodies, so we need to take it in from outside sources. Thousands of Napoleon's troops died during his retreat from Moscow because their wounds could not heal as a result of lack of salt in their diet.

Because we and all other animals have to get salt from the outside in order to stay alive, our behaviors need to contribute to that quest. A behavior that has been experimentally induced, called "salt appetite," can be seen when a laboratory rat is deprived of sodium. What you see is the rat actively searching and sniffing about and taking little nibbles of everything in his environment—*as if looking for something*—and he will continue to hunt in this manner until he finds food that contains salt. When the rat finds a food source containing sodium chloride, his sense of taste innately enables him to recognize the nutrient he needs and he will immediately devour enough of that food to restore his sodium levels to normal. Salt appetite, like that exhibited by sodium-restricted rats, is rarely seen in humans and is of course not ethical to induce. However, extreme conditions, like war, often produce scenarios that can lead to severe sodium deficits and most likely also to some behaviors that are reminiscent of salt-starved rats.

People everywhere like salt, but North Americans have a particularly salty tooth. Although salt consumption is believed to have peaked in Europe in the nineteenth century when people ate as much as 18 grams per day in the form of ham, bacon, and other salted meats and fish, the majority of Americans today also consume upward of 18 grams of salt daily. Adults only need 0.5 grams of salt per day to stay alive, and medical guidelines recommend not exceeding 6 grams. This means that average Americans consume 300 percent more salt each day than they should, and salt overconsumption has been directly linked to health conditions, such as hypertension and cardiovascular disease.

The main reason we eat so much salt is because we like it so much. But it turns out that how much we like it is due to how much of it we eat. In a study conducted in the early 1980s at the Monell Chemical Senses Center, a group of young adults was placed on a self-maintained low-sodium diet for five months. The participants' taste preferences to salt in soup and crackers were recorded two months prior to initiating the reduced-salt diet and then during the five-month period when their sodium intake was lowered. A control group of adults who did not curb their salt intake also provided ratings for their salt preferences in soup and crackers over the same time period for comparison. The results from this study revealed that the people in the salt-restricted group markedly decreased their preferred salt levels in soup and crackers, but the control group participants did not.[4] How much salt we like in our food can be lowered by simply changing our "salting" behavior. In another study, participants were given a salt pill to swallow every day while the amount of salt they *tasted* in food was reduced. That is, the quantity of salt they were ingesting didn't change, but the amount they experienced in their mouth did. These participants also developed a lower preference for the taste of salt in food.[5] This means that by reducing how much salt we "taste" we can effectively diminish our desire for salty food.

It would seem that if we could just bite the bullet and eat fewer potato chips and take the salt shaker off the table, we would end up wanting and eating less salt, but making people adhere to or even initiate salt-reduced diets is difficult because it is such a desirable taste and really does make our food experiences better. Moreover, the amount of salt we add to our food through our own control makes up only 15 percent of the salt we consume; the rest is either found naturally in the foods we eat—a tomato has 14 milligrams of sodium and a cup of milk contains 122 milligrams—or has been added during food processing. Approximately 75 percent of the salt we consume comes from processed foods, even processed products that don't seem categorically salty. An average store-bought white bagel contains all the salt we physiologically need in a day, 500 milligrams; so does a half a cup of cottage cheese. We would have to be intensely scrupulous and spend hours reading labels, or grow, hunt, and prepare all of our food without salt, if we wanted to take charge and effectively minimize the amount of salt we consume.

Considering how much salt there is in the processed food we regularly enjoy, our understanding of the negative health consequences of overconsumption, and our innate and unabating love of salt, there ought to be a gold mine awaiting the artificial salt industry. However, though sweet substitutes have been relatively straightforward to develop, salt substitutes have not. This is because the biochemical mechanisms by which salt taste is experienced are still unknown. Currently, trying to find a nonsodium, safe, salty-tasting chemical is like trying to find a needle in a haystack. Because the receptor, or receptors, for sodium in our tastes buds have not yet been discovered, finding salty taste potentiators the way Senomyx is doing with the sweet receptors is currently impossible. Those in the salt business are hunting feverishly for the answers to this puzzle and you can be sure once they are found you will find the fruits of that labor on the market.

THE JOY OF FOOD

Some people eat to live, others live to eat. No matter what kind of person you are, if you have lost your sense of smell, you can now only eat to live. This is why when people become anosmic, like Michael Hutchence, they often think they have lost their sense of taste as well. Almost always their sense of taste is functionally intact, but their experience of food is so diminished that they assume their taste is gone, too. Without a sense of smell, Sprite and Coke taste exactly the same, a peach tastes of sweet and a bit of sour, and steak is just salty. What we are really talking about when we tell each other about the "taste" of the artichokes we had last night is "flavor." Flavor is the combination of the basic taste sensations with smell.

As you eat, you smell your food twice: once through your nose as food approaches it on the way to your mouth, called *orthonasal olfaction,* and again when the food is in your mouth, called *retronasal olfaction,* when the odors emitted from foods and beverages move from the mouth, back (thus "retro"), and up behind the roof of the mouth into the nasal cavity. Retronasally perceived odors seem to come from "taste" in our mouth, but this is an illusion. The fact that people feel sure that they have lost their sense of taste when they lose their sense of smell demonstrates how strong this illusion is. Our brains trick us by seamlessly knitting together the sensations of taste and smell to give us the flavors that make up our food world.

Like orthonasal olfaction, retronasal olfaction is dependent on airflow, which is why when you have a cold, food "doesn't taste right." When your nose is blocked, you don't get the orthonasal aroma as you bring the golden, crispy French fries to your mouth, and similarly you can't get any retronasal sensation because the airflow from your mouth to your nose is obstructed. So while those golden sticks may look delicious and you have a memory of their pleasurable flavor sensation, when you have a cold, eating French fries is more like eating greasy, salty cardboard. The next time you have a cold, ask someone to blindfold you and test whether you can tell the difference between a raw potato and an apple. Red wine and coffee of the same temperature would also be difficult to tell apart.

Another way to see for yourself how significant smell is to flavor (without waiting for flu season) is this simple "jelly bean test."

Hold your nose with two fingers and put a jelly bean in your mouth. Now chew. All you will taste is sweet. Now release your nostrils and continue to chew. Aha, the lime, licorice, strawberry— whatever flavor-scent molecules have been added—will now surprise you with the "taste" of the jelly bean you are eating.

BETTER LIVING THROUGH CHEMISTRY

In the Middle Ages, food was not particularly palatable. Meat was preserved with salt and then cooked until it was charred. When Marco Polo traveled from Italy to the far reaches of Asia and brought spices back to Europe, he revolutionized both the culinary and social order. Empires rose and fell, religions began, wars were fought, and fortunes were made—all in the name of spice. Spices were a gastronomic extravagance that everyone wanted but only the wealthy and powerful could obtain. A pound of ginger would cost the buyer a sheep. Pepper was worth more than gold, and Arab traders sold it by the peppercorn. Remember that the next time you are refilling your pepper grinder and a dozen corns spill onto the floor.

Today, rather than trekking to Ceylon for cinnamon, or getting your vanilla delivered from Madagascar, the "spice" that dominates the commercial world is composed of chemicals designed by the flavor industry. You may not have realized it but manufactured flavors rule our food experiences. The world market for manufactured flavors is about $8 billion per year. Two-thirds of the diet of the typical American is made up of packaged or processed food—that is, flavor-added food, and without those added flavors, your frozen dinner of "chicken cacciatore" would be essentially inedible. It is also the case that most of these flavors are artificial. But the distinction between natural and synthetic is artificial itself—we just like the connotation of "natural" better and hence believe a product with "natural" ingredients is superior.

To demonstrate how illusory the distinction between natural and artificial is, I conducted an experiment where participants had to either guess or were told that various fragrances were natural or synthetic, regardless of whether they were or not. I found that when someone believed a fragrance was natural, either through their own decision or because they were told so, they thought it was more pleasant than if they were under the impression that the exact same aroma was synthetic. Moreover, when asked to determine which fragrances were natural and which synthetic, the individuals did no better than if they had just flipped a coin to decide. The distinction between natural and synthetic is in our minds and our aesthetics.[6]

As with odors, all flavors are chemicals. A flavor is simply an aroma that you can eat, and therefore the safety and derivation guidelines for manufacturing them are different than when the same aroma goes into soap or perfume. Strawberry *fragrance* and strawberry *flavor* produce the same effect in our nose, but one has passed safety tests for edibility and one hasn't.

In the commercial world of flavor, there is no such thing as "*the* flavor of strawberry" or "*the* flavor of onion." Rather, there are thousands of variants of each flavor and depending on what you want it for, the flavor will be somewhat different. Givaudan, the world's largest flavor and fragrance company, with corporate headquarters in Vernier, Switzerland, and flavor operations in

Cincinnati, Ohio, makes six thousand versions of "strawberry" flavor alone. The "strawberry flavor" that is in candy, yogurt, jelly, and Pop-Tarts is different in each case. The jam in strawberry Pop-Tarts isn't trying to taste like fresh strawberries; it wants to be canonical strawberry Pop-Tart flavor. Givaudan also makes four thousand varieties of orange flavor, three thousand chicken flavors, five thousand beef flavors, and a few thousand butter flavors—just for the North American market. Asia and Europe have thousands more, each specific to their culinary preferences.

From a commercial perspective, the bottom line is that artificial sources are much cheaper than natural ones. If you put twenty tomatoes into a jar of pasta sauce, you'll get a gourmet tomato sauce, but it will be expensive to produce. If you reduce the amount of tomatoes and substitute flavor ingredients, it becomes much more cost-effective. Besides replacing foods, flavors can also substitute for cooking processes. Rather than adding ten different spices and meats and slow cooking for six hours, you can add flavor compounds that elicit these sensations and produce a pasta sauce that "tastes" like Bolognese with negligible time and effort. Artificial flavors are also very concentrated so you don't need much to produce the desired effect. A jar of processed pasta sauce only contains about 1 percent flavor ingredients.

ONE MAN'S MEAT IS ANOTHER MAN'S POISON

There are abundant anecdotal examples of the culturally polarized responses we have to specific foods. Asians consider cheese to be disgusting, but most Westerners love it. The Japanese eat "natto," a fermented soy bean dish, for breakfast but other cultures wouldn't touch it. In fact, there is little cross-cultural consensus on what are considered "normal" food flavors. Fried green tomatoes are as appealing to some southerners as fried scorpions are to some natives of North China, but not likely the other way around.

Nevertheless, if you go to Buenos Aires, Kuala Lumpur, or Paris, you can find McDonald's, Wrigley's gum, Coca-Cola, Starbucks, and Snickers bars. With the global community becoming closer and smaller there are many flavors that are recognized and generally liked around the world, and certain cuisines, like Indian, Thai, and Mexican, are available in most cosmopolitan centers. Still, the vast majority of meals eaten by the vast majority of people are composed of very local flavors. Everyone may like baked doughy things, but the qualities of the baked doughy thing that is liked vary across cultures from sweet to spicy to savory. Meaty dumplings in China, spicy empanadas in Guatemala, and sweet apple fritters in the United States taste very different, though they are all fried doughy things.

The favorite flavors of children also vary widely, even between Europe and the United States. In Europe, peach is a highly liked flavor, but in the United States it is one of the least successful. Grape jelly is the number one purple jelly and juice flavor in the United States but not elsewhere. In the United Kingdom black currant flavor is the purple best seller. And though almost all kids like orange juice, not all children like juices that are orange. German children love carrot juice, but only the progeny of health food junkies are likely to ask for it in America. Different ethnic markets also heighten or hide the essential flavors in various foods. In the United States, the soy flavor in soy-based products is often masked because Americans don't much like the actual taste of soy. But in Asia, flavorings are added to soy products to boost the "soy-ness" because it is so well liked.

FOLLOW YOUR NOSE

Ed is walking the streets of Rome and now finds himself transfixed in front of a trattoria. He hadn't realized he was hungry, but the aroma of sautéed garlic, oregano, and tomato is overwhelming—so is the sweet smell of freshly baking dough wafting from the brick oven he can see through the window. "I have to have some real Italian pizza now," he mutters aloud, as he swings open the restaurant door.

Food aromas often trigger cravings for the food in question even for those who don't consider themselves the craving type. These nasal inducements are such effective motivators of food-purchasing behavior that food retailers routinely take advantage of our weaknesses in places like malls and movie theaters. You may have wondered how it is possible that chocolate chip cookies could be giving off such a vigorous mouthwatering aroma when there is no bakery in sight and instead only a small cookie kiosk in the middle of a sterile mall. It is because aroma delivery devices are pumping the scent of tempting goodies into the air. These scenting devices are commercially available and widely used as customer lures. Almost all of the food aromas you are baited with in public places come from a machine, not a kitchen. However, in the "freshly brewed coffee" industry, there are some retailers who magnify the scents from true brewing coffee.

Chronic dieters are especially susceptible to food aromas as cues to their cravings and will eat even more pizza and cookies than nondieters when pizza or cookie aromas waft by their noses and they can no longer resist the temptation. Food aromas are responsible for whetting our appetites, and conversely when people have lost their sense of smell, they often don't know when or how much to eat. Many people who become anosmic report changes in weight that come from either eating more than they used to, in an effort to regain some lost satisfaction, or not eating as much and losing weight because food is less appealing and also because they may not even remember to eat. Most people don't pay attention to internal physiological signals for eating, which is why many people overeat and some busy and distracted people can even "forget to eat." Having a sense of smell allows food aromas to be powerful appetizers for both the forgetful and greedy among us, and without access to aroma we can suffer. Jessica Ross told me that she now only knew when to eat by looking at the clock or having others tell her.

The fact that food aromas can serve as reminders for mealtimes is especially significant for the elderly. About half the population over the age of eighty has effectively lost its sense of smell, which means that olfactory mealtime signals fail for a large proportion of the population. This translates to the fact that quite a few elderly people may be eating an inadequate amount or eating nutritionally inadequate foods. Malnutrition produces a dark side to our thinking and personality, making people irritable, confused, and forgetful—symptoms that look a lot like dementia. The similarity is so strong that elderly people are sometimes misdiagnosed and treated for dementia when all that is wrong with them is malnutrition due to smell loss. Although smell loss can be an early diagnostic sign for dementias, like Alzheimer's disease, it can also be the reason why completely mentally healthy people can act demented. It is therefore critical that elderly people have olfactory function tests before further diagnoses regarding their mental health are made.

REMEMBER WHEN

Cravings can also be instigated by scents linked to memories and associations with seemingly disconnected relevance. One friend told me that when it smells like summer—"like fresh mowed grass and sunshine"—he craves watermelon. This anecdote illustrates how memory and our associations play a role in our cravings. Another friend told me that while driving to work one morning, a sudden memory of eating Dungeness crab with his uncle in San Francisco as a boy brought on such an intense craving for Dungeness crab that he changed the course of his day to get some.

Elizabeth Loftus, the world's leading psychologist on false and recovered memories, believes that memories of our food experiences may be able to make us eat better and lose weight. In collaboration with researchers at the University of California, Irvine, and the University of Washington, Loftus recently conducted an experiment where college students were convinced that they disliked strawberry ice cream.[7] In the first session of the experiment the students completed various food history, preference, and behavior questionnaires. Inserted into the food history questionnaire was a critical item that stated, "Fell ill after eating strawberry ice cream." Participants rated on an eight-point scale how likely it was that each item had happened to them before the age of ten. On the food preferences questionnaire, they were asked to indicate how much they liked *strawberry ice cream,* among sixty-three other foods, and the behavior questionnaire asked them to indicate how likely they would be to *eat strawberry ice cream* at a party, along with thirty-six other food options.

One week later the students returned to the lab and were given false feedback about their questionnaire responses. They were told that a computer had tailored a specific profile for them, which indicated that as children they disliked spinach, enjoyed pizza, felt happy when a classmate brought sweets to school, and . . . *got sick after eating strawberry ice cream.* The participants were asked to think about the ice-cream memory and to answer questions like: "How old were you?" "Where did it occur?" and "Who were you with?" The students then completed the same questionnaires as they had at the first session concerning food history, preferences, and behavior, and changes in their responses to the *strawberry ice cream* item were measured.

Loftus and her colleagues found that twenty-four out of seventy-one participants went from denying that they had gotten ill after eating strawberry ice cream as a child to believing that they had, even though most of them couldn't recall the exact event. Not only did one-third of the students erroneously believe that they had gotten sick on strawberry ice cream, but on their second pass through the food preference and behavior questionnaires, they rated strawberry ice cream as unpleasant and said they would avoid eating it at a party. In another experiment following very similar procedures, Loftus's group implanted a memory in college students that they "loved asparagus the first time they ate it as a child." Forty percent of the participants later believed this and indicated that they wanted to eat asparagus more often.

Could tricks like this have a significant impact on diet and help people lose weight in the real world? Though not everyone is susceptible to suggestion, our beliefs about what is good or bad are especially manipulable in the world of smell and flavor, and "food aversions," even false ones, are particularly potent because of their biological significance to our survival and health. Therefore, implanting false memories may help curb the appetites of those who tend to eat whatever is in front of them, and it might also help direct people to more healthy snack options. The problem is, could you believe that strawberry ice cream, flourless chocolate cake, or chicken wings were sickness inducing if you really loved them? Or simply if you had frequent experiences eating them? Strawberry ice cream is a relatively uncommon treat compared with chocolate or fried foods. From my own introspection, I think that for craved foods and frequently eaten foods, childhood-based suggestions would be hard to implant and especially hard to maintain. Willpower and the denial of pleasure are still crosses that a dieter must bear.

CHAPTER 8

A WHIFF OF THE FUTURE

The nose of today knows no tomorrow.

—RACHEL HERZ

You walk into an airport and wasps zoom across your luggage to check for explosives. You pass through security and an electronic nose scans your body odor and compares your MHC profile against a database of known terrorists. While you are waiting at the gate for your flight, you call in to work, and your cell phone tells you that you have bad breath. Feeling a bit disturbed by that news, you pop open a gadget and click on your favorite smell to get a whiff of emotional relief. Sound like science fiction? All of these olfactory technologies either currently exist or are in advanced prototype stages. Incredible as it may seem, many more fantastical innovations are currently available and being developed to harness smell in myriad ways to facilitate our lives.

SCENTINELS

Gordon checked the sensors and scanned the computer readouts beside the cold storage unit and nodded to Cliff. "The meat is still fine, we can ship it out tomorrow."

What mysterious apparatus could tell Gordon that meat behind five inches of insulation and steel was safe for consumption? An electronic nose, that's what. Electronic noses, or "e-noses," have existed for about ten years, and since their inception continuous advancements in their capabilities have been made. The inspiration for electronic or artificial noses is the human nose itself. When we recognize a chemical as the smell of rhubarb pie, a particular pattern of activation from multiple and distributed receptors in our nose has been interpreted by our brain as rhubarb pie. Similarly, e-noses typically consist of an array of sensors, each with distinct but overlapping sensitivity to a variety of chemical odors that have been previously specified. The array of sensors creates a pattern that is unique to each chemical it detects. This pattern can be digitized and then compared with known templates to determine a match or signal that something is out of the ordinary. The sensitivity of artificial noses can be ten to one hundred times better than the human nose for detecting many, but not all, compounds.

The advantage of e-noses over traditional chemical-sensing techniques is that e-noses can detect multiple forms of various chemicals, whereas traditional sensors can recognize only one type. Imagine the difference between a camera that can only take pictures of blue objects compared with one that can take pictures of objects in every shade between green and violet. Electronic noses are also cheaper, more accurate, and much less labor intensive than previous-generation sensors.

Currently the biggest market for e-noses is the food industry. Applications for assessing food and beverage quality and control range from cooking-process consistency; detecting "off" odors in packaging materials; grading whisky, wine, and other alcohols; and classifying olive oils. The most critical use for e-noses, however, is for sensing spoilage in products like meat, milk, and mayonnaise, where consumption of rancid food can lead to serious health consequences; e-noses are very good at these jobs. In several experiments, the cleverly named Cyranose-320 was used to analyze the surrounding air from fresh beef strip loins stored at 4° and 10°C and was found to be accurate between 90 and 100 percent of the time when compared against actual microbial counts.[1] E-nose technology has a number of advantages over conventional methods. E-noses are cheaper and more accurate than human panels, they can reduce the amount of analytical chemistry that would otherwise be performed, they are easy to use, can produce immediate results, and are portable.

The way the Cyranose-320 and many other e-noses detect odors is by polymer films that vary in electrical conductivity in the presence of different chemicals. Recently, an innovation in e-nose technology has emerged from the laboratory of Vivek Subramanian, a professor of electrical engineering and computer sciences at the University of California, Berkeley. His e-nose uses an array of transistors made of different organic semiconductor materials. The electronic transistors are made of carbon-based materials as opposed to inorganic materials such as silicon or copper.* Transistors made of various organic compounds respond to specific chemicals differently, so an array of sensors will produce a distinctive pattern for every odor. Transistors are superior to other methods such as polymer films because they are highly sensitive to very slight changes in chemical charge and bond structures. A further advantage to Subramanian's e-nose is that it is very cheap and easy to produce. Organic semiconductors can be made by an extrusion process that uses a modified standard inkjet printer to produce circuits, slashing the cost of production from several hundred dollars to only several dollars. Subramanian predicts that in the not-too-distant future a variant of his device will be so cheap that it will be able to be directly built into soup and tuna fish cans, to indicate precisely when a product is past its safe consumption date.

Outside the food and beverage industry, e-noses are now being put to use in the perfume business to identify counterfeit products; at customs checkpoints to search for banned materials; in the chemical workplace to detect and monitor poisons or toxins; for national security to hunt for chemical weapons; and in medicine to detect disease states ranging from tuberculosis to breast cancer.

---

*"Organic," in chemistry terms, refers to carbon-based molecules, and is so named because it is in all living things. "Inorganic" molecules are in nonliving things.

THE BREATH OF LIFE

Some diseases cause volatile organic compounds (VOCs) indicative of the disease state to be present in expired breath. Aliphatic acids, which smell vinegary, show up in the breath of people with cirrhosis of the liver. Di- and trimethylamine, fishy smells, are found in the breath of those with failing kidneys. And lung-cancer patients exhale a cocktail of alkane and benzene derivatives, which also constitute the primary VOCs that compose "new car smell."

The Breath Research Laboratory of Menssana Research Inc. in Newark and Fort Lee, New Jersey, has developed a breath analyzer device that can detect very early stage lung cancer. Dr. Michael Phillips, the lead researcher on the project believes it can also be used to detect breast cancer and tuberculosis. The patient simply breathes into the device for two minutes so that the VOCs that are in the bloodstream and are indicative of disease states can be captured. The levels of VOCs are then measured using a standard volatile chemical analysis technique called *gas chromatography** and are compared against a sample of room air taken at the same time. The system can detect VOCs with concentrations as low as a few parts per

---

*Gas chromatography–mass spectroscopy, or GCMS, is the combination of two techniques that are used to form a single method of analyzing mixtures of chemicals. Gas chromatography separates the components of a mixture, and mass spectroscopy characterizes each of the components individually. By combining the two techniques, an analytical chemist can both qualitatively and quantitatively evaluate a solution containing a number of different chemicals.

trillion. Early tests on VOCs in the exhaled breath of breast cancer sufferers suggest that this device might be more sensitive and accurate, not to mention a much more comfortable screening method, than mammograms.

VOCs in your breath can also indicate superficial facts, like that you haven't brushed your teeth today, or are drunk. The cell phone that tells the user he needs to reach for the Listerine began as an April Fool's press release issued by Siemens Mobile in Germany in 2004, but the tall tale became an international buzz when it was picked up by a swath of news sources ranging from the BBC to Reuters and Yahoo.com. The truth behind the story is that Siemens's corporate technology division in Munich is developing small sensors that can detect a wide variety of odors—in other words, mini e-noses—which will be able to act as early warning devices for fires or other chemically detectable dangers. The sensors' small size (one millimeter square), ability to recognize target gases in minute quantities, and minimal power requirements mean that they *could* be used in portable devices, such as cell phones. Another possible application is in dashboards or steering wheels, where it could prevent the engine from starting if alcohol were "smelled." These applications have not yet been marketed, and it is not clear whether consumers would want their cell phones insulting them, or their cars refusing to let them drive home after happy hour. But on November 20, 2006, Mothers Against Drunk Driving (MADD) announced a national campaign along with the Department of Transportation to enforce alcohol-sensing devices in the vehicles of convicted drunk drivers. The current device, called "Ignition Interlock" or simply "Interlock," requires such drivers to blow into an instrument that measures alcohol concentration in blood. The vehicle won't start unless the driver's blood alcohol level is below a preset minimum. Interlock is essentially a Breathalyzer in your car that is wired to the ignition. Unlike its purely commercial counterpart, it is likely that Interlock devices will become nationally mandatory for convicted drunk drivers. At the time of MADD's November 2006 announcement, forty-six states either already supported the idea or had similar laws on their books.

THE SMELL OF WAR

Driven by current fears of bioterrorism, several companies are also working on e-noses that will be able to check the air for various forms of chemical weaponry and, if and when contaminants are found, will initiate air filtration and disinfection systems. E-noses could also be used in the "war on terror" by sniffing out suspected or known terrorists by identifying their unique MHC/body-odor profile. In fact, the U.S. military is currently funding projects that promise to lead to the development of such devices. It is certainly possible that some day in the not-too-distant future, airport checkpoints will scan your body odor and then compare it against a database where the body odors/MHC profiles of known terrorists are stored.

Sniffing out terrorists is an e-nose job of tomorrow, but artificial noses are currently being used to find explosives and land mines. Buried land mines kill upward of twenty thousand unsuspecting men, women, and children each year. Their lurking presence also inhibits agriculture and economic development in war-torn zones. Labs at Tufts University, MIT, and the U.S. Army's Defense Advanced Research Projects Agency, along with several companies, have recently unveiled an artificial nose for sensing land mines. In honor of the fact that the most sensitive method for sniffing out explosives is with dogs, one version of these land mine sniffers is called Fido. E-Fido is still undergoing field tests and it may be some time before the army has a working version ready for its troops, but results so far have been encouraging.

THE DOGS AND THE BEES

No matter how good e-noses are at finding mines, they are still no match for dogs who can detect compounds that range in concentration from 10^{-16} to 10^{-18} M/L in air. This is the equivalent of being able to detect a chocolate bar in a city the size of Philadelphia. But there are problems with using dogs, not the least of which is their sensitivity to human emotions and the emotional bonding that takes place between handler and dog. Dogs also weigh enough to set off land mines, and they, like us, get bored, hungry, and distracted, and therefore can be inaccurate or unproductive.

It may seem absurd, but a recently conceived animal alternate with surprising accuracy and few problems is the bee. Bees can detect concentrations of chemicals in minute parts per trillion, sufficient acuity to detect any worrisome chemical. Furthermore, bees don't cause land mines to explode, do not require a handler, are inexpensive to maintain, and can be trained almost instantly.

Like dogs, bees are trained to find explosive material using simple Pavlovian conditioning. For example, the chemical odor of interest such as DNT, a derivative of TNT found in most land mines, is wafted over the bees while they receive a food reward like sugar water, thereby linking the smell of land mines to food. Bees learn this association very quickly and within minutes you have an army of explosive hunting bees, who whenever they smell explosives literally behave like "bees to honey"—clustering and hovering over the location of *interest*.

Jerry Bromenshenk, a research professor at Montana State University, is currently using bees for land mine detection. Once trained, bees are released over the field of concern and their "food"-seeking behavior is observed. One of the limitations of using bees, however, is that their flying patterns are harder to track than a bounding eighty-five-pound German shepherd. However, Lidar, which uses laser light the way radar uses radio waves, is a new technology being worked on to solve this problem. Lidar is transmitted over areas where bees are foraging and creates maps of bee density. A swarm of bees hovering over a specific location would indicate a possible land mine. Field experiments testing the efficacy of the bee-Lidar combo have shown it to have between 97 and 99 percent accuracy, and the ability to find DNT sources in less than an hour.[2]

Is the marriage of bees and Lidar the world's solution to the devastation of hidden land mines? The combination certainly helps, but there are still natural impediments—most significantly the fact that Lidar can't tell the difference between bees and ground vegetation that is at the same height as the swarm. This means that mowing the lawn would be required before searching for suspected land mines in a grassy field, an obvious impossibility. Flowers and other pollen sources are also distracters for bees so only fields that are truly food-free can be tested. Furthermore, bees are fair-weather fliers, which means they can only be deployed during the day and in dry and temperate zones. These problems notwithstanding, the benefits and ease of using bees at least in certain locales make them a definitely worthwhile method of land-mine locating. A fringe benefit is their capacity to revitalize agriculture in war-torn countries by their natural pollination behavior. The British firm Inscentinel Ltd. is currently selling trained bees and minihives, in which the insects' responses to various chemicals can be monitored. In addition to finding land mines, the company says their bee systems can be used to screen for chemicals ranging from explosives to drugs and can even perform food-quality control.

Another flying insect that rarely conjures images of safety is the wasp, and yet wasps are now being recruited as a next line of defense in airport security to find hidden chemical weapons and bombs in luggage and storage containers. The idea of turning to wasps for explosives detection arose from the serendipitous discovery that plants such as corn that are preyed upon by caterpillars give out SOS odors that attract a certain species of wasps to their rescue. The female wasp, lured by this SOS, injects her eggs into the cannibalizing caterpillars producing larvae that eventually kill the caterpillars, hence saving the corn. These wasps can be trained to recognize specific odors very quickly, and they also don't sting humans. Trainers use Pavlovian conditioning and sugar water to teach hungry wasps to respond to a target odor. Training takes less than five minutes and a hive of explosive-seeking wasps is ready to go. Perfect? Not quite. It is a challenge to leash a wasp, and despite reassurances that these wasps are harmless, letting them fly loose around airports would itself raise the level of terror.

Joe Lewis of the U.S. Department of Agriculture, and Glen Rains, an agricultural engineer at the University of Georgia, have devised an ingenious solution. Enter the "wasp hound," a ten-inch-long plastic cylinder made of PVC pipe with a hole at one end and a small fan on the other. Inside the pipe is a web camera that connects to a laptop computer for monitoring the behavior of wasps who are housed in a transparent ventilated tube. When the wasps detect the odor they have learned signals "food," they converge, creating a mass of dark pixels on the computer screen. When there is no "food," the wasps mill about randomly within the capsule. These security wasps can work up to forty-eight hours straight. After they have served their country, they are then released to live out the rest of their two- to three-week life span in the wild. The wasp hound is part of a larger government project to determine if insects, and even reptiles or shellfish, can be recruited for defense work. This project has already resulted in refining the use of bees as land-mine detectors. Glen Rains believes the wasp hound could be in your local airport as early as sometime in 2008.

DOGGIE SCHNAUZER MD

In 1989, Dr. Hywel Williams and his colleague A. C. Pembroke reported a case to the medical journal *Lancet* where a border collie–Doberman mix persistently nudged at a mole on his owner's leg—and on one occasion even tried to bite it off. Although the woman was not concerned about the mole, her dog's constant attention eventually prompted her to see a doctor, and a malignant melanoma was discovered. Twelve years later, Dr. Williams and another physician, John Church, again wrote to *Lancet* about a pet Labrador retriever named Parker who began doggedly nuzzling the pant leg covering his owner's left thigh, where previously a patch of eczema had been diagnosed. The unremitting attention of his dog finally provoked the patient to return to his doctor, and a basal cell carcinoma was discovered at the exact spot of the dog's attention. After the skin cancer was removed, Parker lost interest in his owner's leg.[3]

Skin cancer lesions are superficial, so perhaps it isn't so extraordinary that a dog would behave curiously toward a strange spot of scent. But dog noses turn out to be more than skin deep. In 2005, Drs. James Welsh, Darryl Barton, and Harish Ahuja at the University of Wisconsin Cancer Center in Wausau published a report about the ill-fated case of Jane Doe.[4]* Jane, forty-four years old and in good overall health, had recently gotten a dachshund puppy. A few weeks after settling in, the puppy began to pay inordinate attention to Jane's left underarm and would sit beside her relentlessly sniffing and poking her there. After a month of nonstop nudging, Jane pushed the dog aside to feel where the dog had been nuzzling and discovered a lump. Biopsy confirmed cancer. Jane underwent a mastectomy followed by chemotherapy, and then radiation and tamoxifen.

---

*Note that I have used the name Jane Doe, as no name for the patient was given in the report.

Sadly, however, the cancer had already metastasized and she died a year later.

Man's best friends' amazing sense of smell and apparent ability to sniff out cancer has been dubbed "dognoseis" on Internet chat sites and is rapidly gaining scientific attention. Duane Pickel and his colleagues in Tallahassee, Florida, have conducted one of the few controlled studies to test whether dogs really are the new superdoctors.[5] In their study, a standard schnauzer and a golden retriever were first trained to detect the scent of skin cancer melanoma. The dogs were then tested on seven patients suspected of having melanoma. The schnauzer examined all seven patients and "reported" melanoma in five whose disease was subsequently corroborated by biopsy. A sixth patient whose previous biopsy had failed to reveal melanoma was also "reported" as positive by the dog. This prompted further workup and the dog's diagnosis was confirmed. The golden retriever examined four patients and reached the same conclusion as the schnauzer in every case. The likelihood that the schnauzer would be able to diagnose these six patients correctly by chance is 1 in 10 million.

In another recent study, Dr. Carolyn Willis and colleagues at Amersham Hospital in England conducted an experiment to see whether dogs could be trained to sniff out bladder cancer from the odor of urine.[6] Over seven months, six dogs of varying breeds and ages were trained to discriminate between urine from patients with bladder cancer and urine from patients without bladder cancer. To assess "proof of dog" each member of the canine panel was offered a set of seven urine samples on nine different occasions, of which only one came from a patient with bladder cancer. Overall, the dogs correctly selected the bladder-cancer urine on twenty-two out of fifty-four occasions. This success rate of 41 percent was significantly greater than the 14 percent that could be expected by chance. All of the dogs also responded to one of the "cancer-free" samples as positive. This patient had been examined prior to the study and no tumors had been found. However, the patient's doctor was sufficiently concerned by the dogs' behavior to do further tests, and a tumor in the patient's right kidney was revealed. It is not yet known what chemicals the dogs are specifically detecting that leads them to their bladder-cancer diagnosis. Even more remarkable is that they had to pick out this signature scent from among the hundred of other odors that are in urine.

Despite the extraordinary ability of dogs to be the first herald of cancer, it is unlikely that furry, four-legged doctors donning white lab coats will be sniffing patients anytime soon. Far more likely is that once the signature aromas of various types of cancers can be determined, by analyzing the behavior of dogs and their noses, electronic noses will take over the job of screening and prescreening cancer suspects. E-noses are not as cute and cuddly as canine doctors, but the impracticalities of Rover wending his way through gurneys and sick patients are plentiful.

WAG THE WHALE

In addition to saving us, the exceptional abilities of the canine nose are now being harnessed to protect endangered species—in particular, whales. In the North Atlantic and the Pacific Northwest, dogs are helping conservationists and ecologists save right whales and killer whales, respectively. The dogs help their aquatic cousins by finding their scat, which is then studied to figure out why the whale populations are dwindling. The orca population in the Pacific Northwest has declined 20 percent in the past fifteen years and no one yet knows why. On the East Coast, there are currently only 350 North Atlantic right whales left, and only twelve whale calves are born on average each year. By comparison the South Atlantic right whale population averages about thirty new calves per year.

On both the east and west coasts, environmentalists have blamed the depopulation on changes in the availability of the whale's primary food sources, but there are plenty of other insults that could be destabilizing their life cycle. Noise pollution is one rarely considered menace. Whale-watching tour boats that relentlessly chase after whales for the viewing pleasure of the tourists onboard create enough noise to damage whale sonar. Pollution and pathogens in the water itself could also be threatening the whale population, as could stress from encroaching humans and boats. By studying whale scat, a host of important health information can be gleaned, including quality and constituents of diet, nutritional status, stress level, hormone levels, immune function, and metabolic rates. Hormones in feces can reveal everything from whether a whale is pregnant to if it has been affected by biotoxins, like red tide. Statistics for these factors would then enable specific interventions so that a healthier population balance could be restored.

Before turning to dog noses, biologists and conservationists had to hope for luck, happenstance, and their own eyes and nose to find whale scat. Dogs can locate the scent of whale scat that might be barely detectable by eye, not to mention dissolved in the huge body of water around it, and lead biologists to exact spots where whale scat is located. Fargo, a dog currently used in the North Atlantic, is a purebred rottweiler who can smell samples at least one and a half miles away, even if there are only a few flecks floating on the surface. Fargo was trained by being outfitted with a special harness and life jacket and taught to search for floating jars of scat on the water with toy rewards, his favorite being a yellow tennis ball. According to Roz Rolland, a senior whale scientist with the New England Aquarium in Boston who works with Fargo, one of the biggest challenges at the beginning was to contain Fargo's enthusiasm and keep him from jumping off the boat as soon as he picked up the scent of scat. Now he is a four-legged compass for whale waste. In a recent newspaper interview, Rolland told reporters: "The stronger the scent, the faster his tail wags, and then we steer by his nose."[7] Fargo, who previously worked with another dog, Bob, is now the sole member on staff for the North Atlantic whale project. Currently the Center for Conservation Biology at the University of Washington in Seattle uses eleven scat-detection dogs to track the Pacific Northwest orca population.[8]

DOLLARS AND SCENTS

Jon sat down to reprogram his house. He wanted the new ambience to be a surprise for his wife's return from Hawaii. Their $10 million "smart home" was one of the first in the country, and from a central panel he could control the ambient lighting, temperature, and music experienced in each room. Most recently Jon had added a new dimension to his estate's atmosphere—a fragrance device that released exclusive aromatic blends in various rooms through the ventilation and air-conditioning system throughout the day. Jon switched on the console to prepare the special "scent feelings" he wanted to create. First he set the great room to be filled with his favorite scent, Euphoric, timed for his wife's arrival. Next he synchronized the dining room to be suffused with Black Cashmere for their cocktail hour, and last, with a smile, Jon clocked in Inspiration to perfume their bedroom for nighttime.

Computers and odors are not only partnered in the form of e-noses that can detect cancer, chemical weapons, or spoiled food. They are also wedded through gadgets that can produce smells to alter the environments we live, work, and shop in, and to create more realistic emersions into movies and virtual reality. The business world has recently realized the potential of adding smell to the marketing arsenal. Popular business books like *Brand Sense*[9] now explicitly instruct entrepreneurs to capitalize on the emotional dimensionality and memory associations that fragrances can evoke to produce favorable effects on customers. Unfortunately, the first major consumer-goods product to put an olfactory marketing plan directly into action in the United States ended as soon as it began: On December 4, 2006, the California Milk Processor Board, which brought North America the extremely successful "got milk" campaign, decided to test the effects of aromatic scent strips in five bus shelters in San Francisco.

The scent strips smelled like "just-baked cookies," and it was hoped that the aroma would prompt riders to think that they must *get some milk*. On December 5, 2006, however, the San Francisco Municipal Transportation Agency (MTA) ordered the removal of the scent advertisements, apparently because they had received complaints from bus riders worried that the aroma might not be safe; further, the MTA had not been informed by the marketers that any such manipulation to their shelters was going to be tested. Whether this represents the MTA's resentment over being uninformed, California's health paranoia of public odors that has been previously discussed, or our bias to be wary of unexpected or unknown aromas is not known. Clearly, however, so-called bad garlicky and sulfurous aromas are not the only ones that inspire fear. This example also illustrates the power of context in our interpretation of scents. A bleak bus shelter is not a likely place to be fragrant with cookies and therefore may seem suspicious, but a cookie kiosk in the mall or an elegant resort that perfumes its air does not.

Since the mid-1990s upmarket resorts and retail enterprises have recognized the consumer benefits to be gained from using aromas to enhance their establishments. Indeed, the company that Jon employed to install and develop the fragrances for his home is the same company that has been scenting high-end casinos like the Bellagio in Las Vegas, chic hotels such as the Ritz Carlton in Marina Del Rey, and exclusive spas like Bliss in New York City with signature fragrance blends. Founded by Mark Peltier, a self-taught electronics and ideas wizard, and his wife, Eileen Kenney, the fragrance designer behind the company, AromaSys—based in Lake Elmo, Minnesota—creates unique aromas and dispensing technology that allows specific interior spaces to convey distinctive, personal, and embellishing scent atmospheres. The aroma delivery systems work via vaporization technology and a building's HVAC* system, such that only very small quantities of fragrance materials are needed to achieve a high level of stability and distribution over a given interior space, which can be at least as large as a 100,000-square-foot casino.

The main reason that upscale hotels, casinos, spas, and resorts are turning to aromas to enhance their environments is to create positive and lasting "first impressions" and the sense of being in and experiencing a high-quality environment. The assumption is that a superior first impression will make customers more likely to return and to recommend the property to their friends, which over time should result in economic gain. Retail establishments are also taking advantage of the power of scent. The men's clothing store Thomas Pink in New York City uses an AromaSys scent called Line Dried Linen, Victoria's Secret used Angel Heavenly, and AromaSys has recently created a special fragrance for DeBeers Diamond stores in Beverly Hills and New York City.†

Is there actually any monetary success associated with scenting hospitality venues or retail stores? Peltier says his cus-

---

*HVAC (pronounced either "H-V-A-C" or "H-VAK") is an acronym that stands for "heating, ventilating, and air-conditioning," in other words, a building's climate-control system.

†AromaSys only rarely scents private estates and does not typically contract with retail stores.

tomers in hospitality, though receiving numerous compliments from guests, have not recorded any direct sale changes from the addition of aroma. However, his retail clients say they have seen an increase in the number of both patrons and profits since adding scent and changing nothing else. The assumption is that the store's signature scent reinforces an atmosphere of value and prestige, and its thematic congruence with the items being sold creates a holistic environment that is reinforcing and welcoming. If a scented store is in a mall where other stores are unscented, there is the added benefit of creating curiosity, which along with the allure of a pleasant scent should draw in even more potential customers.

This interpretation for the positive effects of scent on customers is logical. But are there data to support the notion that adding aromas to spaces where people spend money actually increases the amount of money spent? Ann Marie Fiore and her colleagues at Iowa State University created a mock retail environment of women's sleepwear and tested 109 female college students for their purchasing attitudes, intentions, perceptions of product quality, and the price they were willing to pay for satin nightgowns and pajamas.[10] An analysis of the responses showed that adding a pleasant fragrance to the product display led to more positive attitudes toward the sleepwear, more committed purchase intention, higher scores on an eleven-point scale stating "I intend to buy this sleepwear," and a willingness to pay higher prices. But in order for the fragrance to have these effects, it couldn't just be pleasant, it had to be thematically appropriate to the items being sold. Lily of the Valley, which was rated as pleasant and thematically consistent with the sleepwear display, led to the increased positive attitudes in purchasing intent and item quality listed above, but Sea Mist, also rated as pleasant but not perceived as thematically consistent with the sleepwear, did not. Extrapolating to a genuine retail environment, the scent of "clean linen" in an upscale clothing store, such as Thomas Pink, is conceptually congruent with the store's theme and products, and Thomas Pink has claimed increased sales since adding Line Dried Linen to the store ambience.

In addition to thematic congruence, a scent's perceived masculinity or femininity and its correspondence with the items being sold turns out to have a genuine effect at the cash register. In the first field experiment of its kind, Eric Spangenberg of Washington State University and several business school colleagues tested whether gender-congruent scents would influence real shopping behavior in a local clothing store.[11] The store sold both men's and women's clothing in equivalent floor-space sections. On various days, over a two-week period, a scent that was pretested to connote "femininity" (vanilla) was dispersed in the store air, and on alternate days, a scent that connoted "masculinity" (rose maroc) was diffused throughout the store air. The researchers found that when the store smelled like vanilla, women shoppers reported more favorable evaluations of both the store and its clothing. Most important, they actually bought more clothes, spent more money on their purchases, and said they were more likely to visit the store again, compared to when the store smelled like rose maroc. The exact same results were found for male shoppers in the reverse; on the days when the store was scented with the masculine fragrance, men's attitudes toward the store and its merchandise were more positive and they also spent more money, bought more clothes, and said they were more likely to shop at the store again, compared to when the store was scented with vanilla.

This is the first study to demonstrate that real shoppers in a real retail environment can be influenced to reach for their wallets by an ambient scent. This finding is exciting validation for the burgeoning implementation of scent marketing. But buyer and retailer beware. A store that sells more than one kind of merchandise concept, such as clothing for both men and women, or cookware as well as lawn furniture, cannot please all of its customers all of the time, unless the items are widely physically separated and an efficient HVAC system is in place. As you will see, the mixing of various aromas in one space runs the risk of creating an odorific environment that is neither meaningful nor pleasant.

NOT JUST THE SMELL OF POPCORN ANYMORE

Designing aromas and aroma delivery devices to manipulate and enhance our environment is hardly new. The pine tree ornament hanging from the rearview mirror of your neighbor's pickup truck is an odor environment manipulator, as are the air fresheners you spray or plug in at home, the apple-cinnamon-scented candle the real-estate agent has burning when you tour a house for sale, and most recently, odor-releasing CDs, like ScentStories created by Procter & Gamble, that play a sequence of fragrances to alter room ambience.

All of these technologies, including AromaSys systems, operate on the principle of releasing odors into the environment. You perceive the odor until you adapt to it or the odor changes. But rather than simply having an odor constantly "on" as a backdrop, what about timing a series of odors to be turned "on" and "off" simultaneously with specific visual and environmental scenes? In other words, can aromas be used effectively in virtual reality or the cinema?

The film industry has been attempting to include our noses for decades, but so far without much success. Smell-O-Vision, invented by Mike Todd Jr., son of the Hollywood movie mogul Mike Todd and stepson to Elizabeth Taylor, was used in his 1960 film *Scent of Mystery*. Movie theater seats were outfitted with a system that piped smells directly to the viewers' noses. This was the first and last film to use Smell-O-Vision in this form. In the 1980s "scratch and sniff" was experimented with, most notably in John Waters's film romp *Polyester* (1981), starring the famous transvestite Divine. For the "odorama" used in *Polyester,* numbered scratch-and-sniff cards were issued to viewers when they entered the theaters, with instructions to scratch the appropriately numbered circle when it appeared in the corner of the screen. The cards also included warnings "do not scratch until you receive instructions from the film." Odorama left the box office when *Polyester* did.

The latest revival of this technology has emerged in Japan. In the spring of 2006, lucky moviegoers who bought the "premium aroma seats"—the last three rows of one theater in Tokyo, and another in Osaka—experienced a scent-enhanced presentation of Terrence Malick's adventure film, *The New World,* starring

Colin Farrell. The Japanese telecommunications company NTT Communications teamed up with the Japanese film distributor Shochiku and developed a system that in the "premium aroma" zone mixed and released seven different aromas during specific emotional scenes in the film according to a computer-controlled schedule. The love scene was accompanied by a floral scent; a mix of peppermint and rosemary embellished a heartrending moment; for joy, the audience smelled a mix of orange and grapefruit; and a scene where Farrell's character was enraged was accompanied by an herblike concoction.

I am impressed that the film and scent technology people behind aromatizing *The New World* recognized that pairing various smells with emotional scenes would be more effective than trying to match odors with momentary visuals. In earlier incarnations, attempts were made to link scents to specific visual imagery: pipe tobacco and baking bread in *Scent of Mystery* and garbage and old socks in *Polyester*. However, the objects that would have emitted an obvious scent did not do so in *The New World,* and this created its own set of problems. For example, in one scene the Algonquin princess Pocahontas smells the pages of a book, but no corresponding aroma is released, and this absence was reported as acutely felt for the scent-expectant viewers.

BRAVE NEW NOSE

Mainstream virtual reality (VR) and video gaming has not yet been infiltrated by aromas, but the military, which is usually years ahead in the technology that the public will one day enjoy, has been.* The Institute for Creative Technologies (ICT)—in a joint venture with the U.S. military, the University of Southern California, Hollywood, and theme-park designers—is putting scent to work with VR to train soldiers for their first real tours of modern urban warfare. As of 2006 the Department of Defense had already committed $145 million to ICT for the development of VR war simulators.

The smell of burning bodies, exploding bombs, blood, and sewage are part of war, and these scents can be very distracting and disturbing for new recruits. Nevertheless, these scents are part of what realistically defines the modern battlefield. The military believes that by using smell combined with VR simulations, the immersion and the quality of training will be augmented and a generation of better soldiers will be born.

Because smells trigger immediate and intense emotional responses, they can prepare soldiers for the scent realities of death and battle and may also mitigate the future development of scent-evoked posttraumatic stress disorder. The emotional arousal that odors can induce also facilitates learning. In this regard, VR scent technology is being used to teach trainees to recognize specific dangers. For example, the smell of burning wires in a flight simulator could signal an electrical fire, and the smell of cigarette smoke might reveal a hidden enemy soldier.

ICT's aroma-emitting VR device is called the "Scent Collar." Recruits wear the Scent Collar while they watch a combat-zone VR simulation, and at various points in the simulation specific aromas are released. The first Scent Collar prototype

---

*The military was using computers as early as 1941.

was completed in 2002, the second in 2005, and a third—which will have ten scent-emitting modules—is in the pipeline. The current working Scent Collar contains four modules, each housing a reservoir with a fragrance-soaked wick. Within each module a small arm can move or open ports to release a controllable amount of scent into the chamber above. A fan helps control the amount of scent that reaches the recruit's nose, as well as odor delivery and clearance speed.

DO ANDROIDS DREAM OF SCENTED SHEEP?

Beyond prefabricated scents for specific visual scenes, like smelling burning rubber as you watch the Humvee catch fire, the dream for scent technology is that from a predetermined palette of aromas, the entire world of odors can be created and implemented in various technologies. Echoing earlier inventions such as DigiScents in the 1990s, the Tokyo Institute of Technology has just invented a device that can apparently digitally record and then re-create any aroma. The current model of this device has ninety-six chemicals in its odor lexicon and fifteen sensors. The sensors are primed to recognize a wide range of odors. When an odor is recognized, a digital algorithm is generated so that the aroma can then be re-created from the ninety-six-chemical vocabulary stored in the device. The expectation is that when the device is retooled to be small enough to be portable, it can be implemented into cell phones and digital recording devices. An olfactorily digitized cell phone of the future will be able to take a picture of the beach you are vacationing on, tell you that you've had too many daiquiris, and capture the scent of suntan lotion and sea air and then re-create it for your friends when you call them on their cell phones to gloat about the good time you're having.

These ingenious scent-computerized and VR devices would seem to open an almost limitless world of commercial possibilities. But there remain fundamental problems with successful implementation of these innovative technologies due to our incomplete understanding of how the sense of smell actually works, and several inherent incompatibilities between the nature of vision and smell.

The concept of digital odor remastering from a palette of chemicals that is the basis for the Japanese device implies that there are "odor primaries." In vision, the light primaries of color are red, green, and blue. Lights of red, green, and blue wavelengths can be mixed together to produce all visible color. Researchers in olfaction have been searching for a set of so-called odor primaries for decades, but no set has ever been determined. Moreover, unlike vision, olfaction is not a simple synthetic sense, where a new holistic sensation arises out of a combination of other sensations. You can mix red and blue to make a new color, purple, but you can't necessarily mix a chemical that smells like grass with one that smells like coffee and get a new scent "graffee," with its own unique olfactory sensation—it may just smell like grassy coffee.

The relationship between fragrance molecule and fragrance perception is not predictable. Chemicals with very different molecular structures can smell nearly indistinguishable, and molecules with nearly identical structures can smell entirely different. Simply reversing the molecular spin, as in the right-

and left-hand isomers of the molecule "carvone," yields one that smells like spearmint and another that smells like caraway. Because of these random relationships, trial and error is the rule in the commercial world of creating new odor molecules. Fragrance manufacturers typically synthesize a thousand new molecules to get one that they can use. These problems reflect the current mysteries in olfaction and imply that digital recording and resynthesizing devices are unlikely to succeed, at least until we know more. Takamichi Nakamoto, the creator of the odor digital recorder, told me he picked 96 chemicals, instead of 76 or 102, simply because it was the maximum number his device could hold. He also admitted that this set could not re-create all odors, and that what number and what set of odors it could create was still being worked on.

The second set of problems with scented VR stems from the intrinsic nature of the sense of smell and its differences with vision. For odors to work together with film or VR a specific scent has to be simultaneous with a specific visual image. A scent also has to disappear once the scene has changed, and the various aromas emitted must not contaminate one another. However, one of the most fundamental physical challenges for Smell-O-Vision is that olfaction is a very slow sense. It typically takes at least four hundred milliseconds after an odor has been presented for a smell sensation to be detected. By the time you say "aha, rose," the chemical for rose has been in the air for almost half a second. By contrast it takes your brain only forty-five milliseconds, one-tenth the time, to register that you've seen something. Matching fast-paced visual drama with scents is therefore functionally limited by the slowness of smell.

Besides taking a long time to turn olfaction on, it also takes a long time to turn it off. Airflow, temperature, humidity, volume, and a host of other factors determine how long a scent will persist in your nasal panorama. Another related problem is odor mixing. The movie scene changes from a dockyard to a graveyard and the scent of fish and brine mixes with someone's creative idea of damp earth and bones. What happens when the next cutaway is to mobsters eating in an Italian restaurant? Not a very good mélange and, more important, not an accurate one. Even when smells are emitted sporadically over the course of a feature-length film, airflow and mixing problems can occur. One viewer of *The New World* in Japan, where only seven odors were used, described his experience as: "Like watching a movie while an aromatherapy clinic was being held in the lobby. Even in my Premium Aroma Seat, I had a hard time distinguishing the scents and often was unsure if a new perfume were being introduced or if a random atmospheric shift had brought a residual scent into stronger focus."[12]

It is also the case that molecules we can smell are chemically sticky; they literally stick to paint, cloth, and plastics—things that movie theaters are filled with. Moreover, most aromas used in scenting devices are in an oil base, which is sticky itself. After one showing of *The New World,* the cinema would be semipermanently coated with the aromas from the film, though it might not be noticeable until it had been showing for a week or so. The next Smell-O-Vision film to come along would have to compete with the previous film's aromas. A few such films and the theater would be a strange olfactory brew indeed. It is hardly feasible to repaint or reupholster a movie theater every time a film has run its course, yet this issue of contamination is one of the problems that has to be dealt with if Smell-O-Vision can ever be a commercial reality. ICT's Scent Collar has the advantage of a small surface area and emits only limited amounts of precisely targeted scent. It also has a fan to facilitate scent arrival and departure. Compared with theater-scale versions, the problems of contamination and speed of dispersal and clearance are reduced, but not eliminated.

Finally, there is the olfactory thorn of adaptation discussed earlier. We adapt to and can no longer "smell" a scent that is constant in our environment. Repeated, intermittent exposure to the same smell will also eventually produce adaptation. This may not be a major problem for the film industry, unless producers are concerned about their obsessive fans who need to see a movie dozens of times. However, this is a practical concern for VR games that want to include an odorific dimension. If the same scents are emitted every time the game is played, the olfactory component will become irrelevant over time. Similarly, when training soldiers by using Scent Collars, the device should not be overused. Too many mock missions with the same set of smells will result in diminished ability to get the real feel of "the enemy."

A VISION FOR TOMORROW'S NOSE

There are vast uncharted areas where our lives can be enormously benefited by innovations involving the sense of smell and scents themselves. The dreams and realities of diverse olfactory technologies, from diverse sources and toward diverse applications, is phenomenal. However, I think that simple aware-ness of how amazing, wonderful, and incredible our sense of smell is, and how much pleasure, dimensionality, intensity, and meaning it can bring to our lives, is the most essential olfactory knowledge that we need to enrich our lives now and in the fu-ture. Paying attention to smells really does enhance our ability to smell, and if we don't pay attention, many aromas will slip by unappreciated.[13]

I am not snubbing the possibilities of digital technologies, the fusions of smell with our other sensory experiences, or har-nessing wasp and dog noses, but since I value our sense of smell and our human experience of smell so much, what I would most like to see are technologies that can help people who have been deprived of the joy of smell. Though currently at the level of fantasy, I believe it is not too whimsical to imag-ine "olfact-aides" as a viable innovation of the future.

As indicated earlier, during normal aging, our sense of smell, like our other faculties, declines in acuity. This seems primarily due to the fact that the ability to regenerate new olfactory re-ceptors becomes impaired, so that the cell-death to cell-regeneration equation becomes unbalanced. The more functioning receptors we have, the stronger a scent sensation is. This is why smokers often complain that they "can't smell very well." They are right. The toxins in cigarette smoke kill off olfactory receptors. However, because we normally regenerate olfactory receptors every twenty-eight days, a month after quit-ting, a former smoker will be able to smell the roses as well as anyone else. During aging, however, the ability to regenerate olfactory receptors falls off without regain. If a twenty-year-old with a keen sense of smell has ten thousand functioning receptors, an eighty-five-year-old who can't smell the difference between standing in Starbucks and standing in a butcher shop has a much smaller number of functioning receptors.

My dream is for a topical solution—a gel, for example, that could be applied inside the nose to substantially boost the gain on any functioning olfactory receptors, such that a little stimulation could go a long way. In other words, with relatively few functioning receptors plus this future medication, the available receptors could produce as much or nearly as much "smell" sensation as healthy full stimulation. Olfact-aide gel would therefore amplify smell sensations for the olfactorily impaired, so that your grandmother will know that a Cinnabon kiosk is nearby without looking. Indeed, anyone at any age who has an impaired sense of smell due to olfactory receptor damage could be benefited by this invention.

That this dream could be close to reality is encouraged by companies like Patus, in Israel, which are already involved in developing ointments that affect the normal functioning of olfactory receptors. OdorScreen, as discussed in Chapter 6, is a topical gel that protects rescue workers from smelling offensive scents that are nonetheless there, by interfering with olfactory receptor activity. Though this application turns olfactory sensation off or down, reversing to the positive is not implausible and is already an underlying method in artificial-sweetener technology. Senomyx, as discussed in Chapter 7, is developing "sweet taste potentiators" that boost the gain in our perception of sweetness, such that a little bit of sugar plus potentiator produces the same sweetness sensation as a big dose of sugar naturally would. Similarly, olfact-aide gel would contain some kind of odor receptor potentiator, so that a little stimulation would "smell" like a lot.

The solution to which chemicals could do this job might come from innovations that are currently applied in the pharmaceutical industry in their search for new drugs. Olfactory receptors are in the same family of receptors that most druggable compounds work through, known as *G-protein coupled receptors (GPCRs)*. Traditional methods of drug discovery, and the lucky observations that gave companies like Pfizer Viagra, are rapidly exhausting the well of possible blockbusters, and novel approaches to finding medications are now needed. The pharmaceutical industry is very interested in maximizing the therapeutic utility of GPCRs because so many successful drugs act through these mechanisms. Rather than looking for more needles in the haystack, a new approach being used is to develop drugs that increase the efficacy of known GPCRs, thereby increasing the potency of natural biochemical interactions. There is a direct analogy between this methodology and how appropriate chemicals for olfact-aide gel could be found, and even the possibility that the specific outcomes from pharmaceutical research, because they involve GPCRs, could directly indicate what chemicals would be best suited for such a product. The sweetness potentiators created by Senomyx work on the same principles, and sweet taste is also mediated by GPCRs.

From burgeoning multidisciplinary interest and approach, and hence minds and money, I believe that olfact-aides are a reality for the not-too-distant future. Olfact-aides may not be able to help people like Michael Hutchence or Jessica Ross who have lost their sense of smell in their brain rather than their nose, but for all of us who live to see healthy old age, and for anyone who has peripheral olfactory damage, they could have tremendous beneficial impact.

Our sense of smell is a treasure we often take for granted and my hope is that without the devastating lesson of its loss, this book has helped you to engage it and appreciate it. Recognizing how essential scent is to our humanity—emotionally, physically, sexually, and socially—and how the experience of scent enriches, improves, and deepens our lives in multiplicative and multifaceted ways gives remarkable meaning to our lives. Our sense of smell enables us to know ourselves and influences our sociability with others. Our sense of smell facilitates our ability to learn and to remember and can alter our behavior. Our sense of smell allows us to experience an intense emotional life, awakens our memories, is interlaced with our mental health, and triggers our passions. Our sense of smell even tells us with whom it would be biologically best to conceive a child, and it can make someone with movie-star looks unappealing and the plainest person the object of obsessive passion. Our sense of smell is truly our sense of desire.

NOTES

PREFACE

1. Ehrlichman, H., & Halpern, J.N. (1988). Affect and memory: Effects of pleasant and unpleasant odors on retrieval of happy and unhappy memories. *Journal of Personality and Social Psychology,* 55, 769–779.

CHAPTER I

1. As described by Mike Gee. (1998). *The final days of Michael Hutchence.* London: Omnibus Press.
2. For further reference, see Deems, D.A., Doty, R.L., Settle, R.G., Moore-Gillon, V., Shaman, P., Mester, A.F., Kimmelman, C.P., Brightman, V.J., & Snow, J.B., Jr. (1991). Smell and taste disorders: A study of 750 patients from the University of Pennsylvania Smell and Taste Center. *Archives of Otolaryngology, Head and Neck Surgery, 117,* 519–528; Takaki, M., Furukawa, M., Tsukantani, T., Costanzo, R.M., DiNardo, L.J., & Reiter, E.R. (2001). Impact of olfactory impairment on quality of life and disability. *Archives of Otolaryngology, Head and Neck Surgery, 127,* 497–503.
3. American Psychiatric Association. (1994). *The diagnostic and statistical manual of mental disorders* (4th ed.). Washington, DC: Author.
4. Pause, B.M., Miranda, A., Goder, R., Aldenhoff, J.B., & Ferstl, R. (2001). Reduced olfactory performance in patients with major depression. *Journal of Psychiatric Research, 35,* 271–277.

5. Zelano, C., Bensafi, M., Porter, J., Mainland, J., Johnson, B., Bremner, E., Telles, C., Khan, R., & Sobel, N. (2005). Attentional modulation in human primary olfactory cortex. *Nature Neuroscience, 8,* 114–120.

6. Chen, D., & Haviland-Jones, J. (2000). Human olfactory communication of emotion. *Perceptual and Motor Skills, 91,* 771–781.

7. Buck, L., & Axel, R. (1991). A novel multigene family may encode odorant receptors: A molecular basis for odor recognition. *Cell, 65,* 175–187.

CHAPTER 2

1. Mennella, J.A., & Garcia, P.L. (2000). Children's hedonic response to the smell of alcohol: Effects of parental drinking habits. *Alcoholism: Clinical and Experimental Research, 24,* 1167–1171; Mennella, J.A., Jagnow, C.P., & Beauchamp, G.K. (2001). Prenatal and postnatal flavor learning by human infants. *Pediatrics, 107,* E88.

2. Haller, R., Rummel, C., Henneberg, S., Pollmer, U., & Koster, E.P. (1999). The influence of early experience with vanillin on food preference in later life. *Chemical Senses, 24,* 465–567.

3. Ackerman, D. (1990). *A natural history of the senses.* New York: Random House

4. Moncreiff, R.W. (1966). *Odour preferences.* New York: Wiley.

5. Cain, W.S., & Johnson, F., Jr. (1978). Lability of odor pleasantness: Influence of mere exposure. *Perception, 7,* 459–465.

6. Herz, R.S., Beland, S.L., & Hellerstein, M. (2004). Changing odor hedonic perception through emotional associations in humans. *International Journal of Comparative Psychology, 17,* 315–339.

7. Robin, O., Alaoui-Ismaili, O., Dittmar, A., & Vernet-Mauri, E. (1998). Emotional responses evoked by dental odors: An evaluation from autonomic parameters. *Journal of Dental Research, 77,* 1638–1946.

8. Suskind, P. (1986). *Perfume.* New York: Alfred A. Knopf.

9. For further information see: Chen, D. & Dalton, P. (2005). The effect of emotion and personality on olfactory perception. *Chemical Senses, 30,* 345–351.

10. Twain, M. (1985). *The signet classic book of Mark Twain's short stories.* New York: New American Library (Penguin Putnam).

11. Herz, R.S., & von Clef, J. (2001). The influence of verbal labeling on the perception of odors: Evidence for olfactory illusions? *Perception, 30,* 381–391.

12. Van Toller, C., Kirk-Smith, M.D., Wood, N., Lombard, J., & Dodd, G.H. (1983). Skin conductance and subjective assessments associated with the odour of 5-alpha-androstan-3-one. *Journal of Biological Psychology, 16,* 85–107.

13. Classen, C., Howes, D., & Synnott, A. (1994). *Aroma: The cultural history of smell.* New York: Routledge.

CHAPTER 3

1. Proust, M. (1928). *Swann's way.* New York: The Modern Library, p. 62.

2. Laird, D.A. (1935). What can you do with your nose? *Scientific Monthly, 41,* 126–130.

3. Rubin, D.C., Groth, E., & Goldsmith, D.J. (1984). Olfactory cuing of autobiographical memory. *American Journal of Psychology, 97,* 493–507.

4. Herz, R.S. (1998). Are odors the best cues to memory? A cross-modal comparison of associative memory stimuli. *Annals of the New York Academy of Sciences, 855,* 670–674.

5. From Proust, M. (1928). *Swann's way.* New York: The Modern Library, p. 65.

6. Herz, R.S. (1997). The effects of cue distinctiveness on odor-based context-dependent memory. *Memory & Cognition, 25,* 375–380.

7. Herz, R.S. (1997). Emotion experienced during encoding enhances odor retrieval cue effectiveness. *American Journal of Psychology, 110*, 489–505.

CHAPTER 4

1. Lehrner, J., Marwinski, G., Lehr, S., Johren, P., & Deecke, L. (2005). Ambient odors of orange and lavender reduce anxiety and improve mood in a dental office. *Physiology & Behavior, 86*, 92–95.

2. Villemure, C., Slotnick, B.M., & Bushnell, M.C. (2003). Effects of odors on pain perception: Deciphering the roles of emotion and attention. *Pain, 106*, 101–108.

3. Gedney, J.J., Glover, T.L., & Fillingim, R.B. (2004). Sensory and affective pain discrimination after inhalation of essential oils. *Psychosomatic Medicine, 66*, 599–606.

4. Campenni, C.E., Crawley, E.J., & Meier, M.E. (2004). Role of suggestion in odor-induced mood change. *Psychological Reports, 94*, 1127–1136.

5. Slosson, E.E. (1899). A lecture experiment in hallucinations. *Psychological Review, 6*, 407–408.

6. O'Mahoney, M. (1978). Smell illusions and suggestions: Reports of smells contingent on tones played on television and radio. *Chemical Senses and Flavor, 3*, 183–189.

7. Dalton, P. (1999). Cognitive influences on health symptoms from acute chemical exposure. *Health Psychology, 18*, 579–590.

8. Das-Munshi, J., Rubin, G.T., & Wessely, S. (2006). Multiple chemical sensitivities: A systematic review of provocation studies. *Journal of Allergy and Clinical Immunology, 118*, 1257–1264.

9. Bornschein, S., Hausteiner, C., Zilker, T., & Forstl, H. (2002). Psychiatric and somatic disorders and multiple chemical sensitivity (MCS) in 264 "environmental patients." *Psychological Medicine, 32*, 1387–1394.

10. Lax, M.B., & Henneberger, P.K. (1995). Patients with multiple chemical sensitivities in an occupational health clinic: Presentation and follow-up. *Archives of Environmental Health, 50,* 425–431.

11. Epple, G., & Herz, R.S. (1999). Ambient odors associated to failure influence cognitive performance in children. *Developmental Psychobiology, 35,* 103–107.

12. Herz, R.S., Schankler, C., & Beland, S. (2004). Olfaction, emotion and associative learning: Effects on motivated behavior. *Motivation and Emotion, 28,* 363–383.

13. Van den Bergh, O., Devriese, S., Winters, W., Veulemans H., Nemery, B., Eelen, P., & Van de Woestijne, K.P. (2001). Acquiring symptoms in response to odors: A learning perspective on multiple chemical sensitivity. *Annals of the New York Academy of Sciences, 933,* 278–290.

CHAPTER 5

1. Daly, M., & Wilson, M. (1983). *Sex, evolution and behavior* (2nd ed.). Boston: Willard Grant Press.

2. Miller, G.F. (1998). How mate choice shaped human nature: A review of sexual selection and human evolution. In C. Crawford & D. Krebs (Eds.), *Handbook of evolutionary psychology: Ideas, issues, and applications* (pp. 87–130). Mahwah, NJ: Lawrence Erlbaum.

3. Daly, M., & Wilson, M. (1988). *Homicide.* New York: Aldine de Gruyter.

4. Buss, D.M., Larsen, R.J., Westen, D., & Semmelroth, J. (1992). Sex differences in jealousy: Evolution, physiology and psychology. *Psychological Science, 3,* 251–255.

5. For the first reported evidence of the MHC-body-odor genetic connection in rodents and humans, respectively, see Boyse, E.A., Beauchamp, G.K., Yamazaki, K. (1987). The genetics of body

scent. *Trends in Genetics, 3,* 97–102; Wedekind, C., Seebeck, T., Bettens, F., and Paepke, A.J. (1995). MHC-dependent mate preferences in humans. *Proceedings of the Royal Society of London Series B, 260,* 245–249.

6. Ober, C., Weitkamp, L.R., Cox, N., Dytch, H., Kostyu, D., & Elias, S. (1997). HLA and mate choice in humans. *American Journal of Human Genetics, 61,* 497–504.

7. Wedekind, C., Seebeck, T., Bettens, F., & Paepke, A.J. (1995). MHC-dependent mate preferences in humans. *Proceedings of the Royal Society of London Series B, 260,* 245–249.

8. Daly, M., & Wilson, M. (1988). *Homicide.* New York: Aldine de Gruyter.

9. Penn, D., & Potts, W. (1998). MHC-disassortative mating preferences reversed by cross-fostering. *Proceedings of the Royal Society of London Series B, 265,* 1299–1306.

10. Martins, Y., Preti, G., Crabtree, C.R., Runyan, T., Vainius, A.A., & Wysocki C.J. (2005). Preference for human body odors is influenced by gender and sexual orientation. *Psychological Science, 16,* 694–701.

11. Milinski, M., & Wedekind, C. (2001). Evidence for MHC-correlated perfume preferences in humans. *Behavioral Ecology, 12,* 140–149.

12. McBurney, D.H., Shoup, M.L., & Streeter, S.A. (2006). Olfactory comfort: Smelling a partner's clothing during periods of separation. *Journal of Applied Social Psychology, 36,* 2325–2335; Shoup, M.L., Streeter, S.A., & McBurney, D.H. (in press). Olfactory comfort and attachment within relationships. *Journal of Applied Social Psychology.*

13. Grammer, K., & Thornhill, R. (1994). Human (Homo sapiens) facial attractiveness and sexual selection: The role of symmetry and averageness. *Journal of Comparative Psychology, 108,* 233–242.

14. Henberger, E., Redhammer, S., & Buchbauer, G. (2004). Transdermal absorption of (−)tinalool induces autonomic deactivation but has no impact on ratings of well-being in humans. *Neuropsychopharmacology, 29,* 1925–1932.

15. Preti G., Wysocki, C.J., Barnhart, K.T., Sondheimer, S.J., & Leyden, J.J. (2003). Male axillary extracts contain pheromones that affect pulsatile secretion of luteinizing hormone and mood in women recipients. *Biology of Reproduction, 68,* 2107–2113.

CHAPTER 6

1. Sullivan, R.M., & Toubas, P. (1998). Clinical usefulness of maternal odor in newborns: Soothing and feeding preparatory responses. *Biology of the Neonate, 74,* 402–408.

2. Porter, R.H., Cernoch, J.M., & McLaughlin, F.J. (1983). Maternal recognition of neonates through olfactory cues. *Physiology & Behavior, 30,* 151–154.

3. Kaitz, M., Goode, A., Rokem, A.M., & Eidelman, A.I. (1987). Mothers' recognition of their newborns by olfactory cues. *Developmental Psychobiology, 20,* 587–591.

4. Platek, S.M., Burch, R.L., & Gallup, G.G. (2001). Sex differences in olfactory self-recognition. *Physiology & Behavior, 73,* 635–640.

5. Porter, R.H., Cernoch, J.M., & Balogh, R.D. (1985). Odor signature and kin recognition. *Physiology & Behavior, 34,* 445–448.

6. Porter, R.H., Balough, R.D., Cernoch, J.M., & Franchi, C. (1986). Recognition of kin through characteristic body-odor. *Chemical Senses, 11,* 389–395.

7. Porter, R.H. (1999). Olfaction and human kin recognition. *Genetica, 104,* 259–263.

8. Chen, D., & Haviland-Jones, J. (1999). Rapid mood change and human odors. *Physiology & Behavior, 68,* 241–250.

9. Haze, S., Gozu, Y., Nakamura, S., Kohno, Y., Sawano, K., Ohta, H., & Yamazaki, K. (2001). 2-Nonenal newly found in human body-odor tends to increase with aging. *Journal of Investigative Dermatology, 116,* 520–524.

10. Corbin, A. (1986). *The foul and the fragrant.* Massachusetts: Harvard University Press.

11. Dollard, J. (1937). *Caste and class in a southern town.* New York: Anchor Books.

12. Hyde, A. (2006). Offensive bodies. In J. Drobnick (Ed.), *The smell culture reader.* New York: Berg, p. 55.

13. Orwell, G. (1937). *The road to Wigan Pier.* Victor Gollanz: London.

14. Classen, C., Howes, D., & Synnott, A. (1994). *Aroma: The cultural history of smell.* New York: Routledge.

15. Marchand, R. (1985). *Advertising and the American dream: Making way for modernity, 1920–1940.* Berkeley: University of California Press.

16. Coates, J.D., Cole, K.A., Michaelidou, U., Patrick, J., McInerney, M.J., & Achenbach, L.A. (2005). Biological control of hog waste odor through stimulated microbial Fe(III) reduction. *Applied and Environmental Microbiology, 71,* 4728–4735.

17. Aftel, M., & Patterson, D. (2001). *Aroma: The magic of essential oils in food and fragrance.* New York: Artisan.

CHAPTER 7

1. Pelchat, M.L., Johnson, A., Chan, R., Valdez, J., & Ragland, J.D. (2004). Images of desire: Food-craving activation during fMRI. *NeuroImage, 23,* 1486–1493.

2. Basson, M.D., Bartoshuk, L.M., Dichello, S.Z., Panzini, L., Weiffenback, J.M., & Duffy, V.B. (2005). Association between 6-n-propylthiouracil (PROP) bitterness and colonic neoplasms. *Digestive Diseases and Sciences, 50,* 483–489.

3. Milunicova, A., Jandova, A., Laurova, L., Novotna, J., & Skoda, V. (1969). Hereditary blood and serum types, PTC test and level of the fifth fraction of serum lactatedehydrogenase in females with gynecological cancer (II. Communication). *Neoplasma, 16,* 311–316.

4. Bertino, M., Beauchamp, G.K., & Engelman, K. (1982). Long-term reduction in dietary sodium alters the taste of salt. *American Journal of Clinical Nutrition, 36,* 1134–1144.

5. Bertino, M., Beauchamp, G.K., & Engelman K. (1986). Increasing dietary salt alters salt taste preference. *Physiology & Behavior, 38,* 203–213.

6. Herz, R.S. (2003). The effect of verbal context in olfactory perception. *Journal of Experimental Psychology: General, 132,* 595–606.

7. Bernstein, D.M., Laney, C., Morris, E.K., & Loftus, E.F. (2005). False beliefs about fattening foods can have healthy consequences. *Proceedings of the National Academy of Sciences, 102,* 13724–13731.

CHAPTER 8

1. Panigrahi, S., Balasubramanian, S., Gu, H., Logue, C., & Marchello, M. (2006). Neural network integrated electronic nose system for identification of spoiled beef. *LWT-Food Science and Technology, 39,* 135–145.

2. Shaw, J.A., Bromenshenk, J.J., & Churnside, J.H. (2005). Polarization Lidar measurements of honey bees in flight for locating land mines. *Optics Express, 13,* 112–121.

3. Williams, H., & Pembroke, A. (1989) Sniffer dogs in the melanoma clinic? *Lancet, 1* (8640), 734; Church, J., & Williams, H. (2001). Another sniffer dog for the clinic? *Lancet, 358* (9285), 930.

4. Welsh, J.S., Barton, S., & Ahuja, H. (2005). A case of breast cancer detected by a pet dog. *Community Oncology, July/August,* 324–326.

5. Pickel, D., Manucy, G.P., Walker, D.B., Hall, S.B., & Walker, J.C. (2004). Evidence for canine olfactory detection of melanoma. *Applied Animal Behavior Science, 89*, 107–116.

6. Willis, C.M., Church, S.M., Guest, C.M., Cook, W.A., McCarthy, N., Bransbury, A.J., Church, M.R.T., & Church, J.C.T. (2004). Olfactory detection of human bladder cancer by dogs: Proof of principle study. *British Medical Journal, 329*, 712–717.

7. From *The Globe and Mail*. Retrieved August 7, 2006, from http://www.theglobeandmail.com/servlet/story/LAC.20060807.WHALE07/TPStory/?query=canine+detective.

8. Weir, K. (August, 2006). Dog chases whale scat. *The Scientist, 20*–21.

9. Lindstrom, M. (2005). *Brand sense*. New York: Free Press.

10. Fiore, A.M., Yah, X., & Yoh, E. (2000). Effects of product display and environmental fragrancing on approach responses and pleasurable experiences. *Psychology & Marketing, 17*, 27–54.

11. Spangenberg, E.R., Sprott, D.E., Grohmann, B., & Tracy, D.L. (2006). Gender-congruent ambient scent influences on approach and avoidance behaviors in a retail store. *Journal of Business Research, 59*, 1281–1287.

12. Fugiwara, C. (July/August 2006). Wake up and smell the new work: An odorama revival has activated noses in Tokyo. Film Society of Lincoln Center. Retrieved August 17, 2006, from http://www.filmlinc.com.

13. Zelano, C., Bensafi, M., Porter, J., Mainland, J., Johnson, B., Bremner, E., Telles, C., Khan, R., & Sobel, N. (2005). Attentional modulation in human primary olfactory cortex. *Nature Neuroscience, 8*, 114–120.

INDEX

Page numbers in *italics* refer to illustrations.